N. Nedjah, L. M. Mourelle (Eds.)

Evolvable Machines

Studies in Fuzziness and Soft Computing, Volume 161

Editor-in-chief
Prof. Janusz Kacprzyk
Systems Research Institute
Polish Academy of Sciences
ul. Newelska 6
01-447 Warsaw
Poland
E-mail: kacprzyk@ibspan.waw.pl

Further volumes of this series can be found on our homepage: springeronline.com

Vol. 144. Z. Sun, G.R. Finnie
Intelligent Techniques in E-Commerce, 2004
ISBN 3-540-20518-7

Vol. 145. J. Gil-Aluja
Fuzzy Sets in the Management of Uncertainty, 2004
ISBN 3-540-20341-9

Vol. 146. J.A. Gámez, S. Moral, A. Salmerón (Eds.)
Advances in Bayesian Networks, 2004
ISBN 3-540-20876-3

Vol. 147. K. Watanabe, M.M.A. Hashem
New Algorithms and their Applications to Evolutionary Robots, 2004
ISBN 3-540-20901-8

Vol. 148. C. Martin-Vide, V. Mitrana, G. Păun (Eds.)
Formal Languages and Applications, 2004
ISBN 3-540-20907-7

Vol. 149. J.J. Buckley
Fuzzy Statistics, 2004
ISBN 3-540-21084-9

Vol. 150. L. Bull (Ed.)
Applications of Learning Classifier Systems, 2004
ISBN 3-540-21109-8

Vol. 151. T. Kowalczyk, E. Pleszczyńska, F. Ruland (Eds.)
Grade Models and Methods for Data Analysis, 2004
ISBN 3-540-21120-9

Vol. 152. J. Rajapakse, L. Wang (Eds.)
Neural Information Processing: Research and Development, 2004
ISBN 3-540-21123-3

Vol. 153. J. Fulcher, L.C. Jain (Eds.)
Applied Intelligent Systems, 2004
ISBN 3-540-21153-5

Vol. 154. B. Liu
Uncertainty Theory, 2004
ISBN 3-540-21333-3

Vol. 155. G. Resconi, J.L. Jain
Intelligent Agents, 2004
ISBN 3-540-22003-8

Vol. 156. R. Tadeusiewicz, M.R. Ogiela
Medical Image Understanding Technology, 2004
ISBN 3-540-21985-4

Vol. 157. R.A. Aliev, F. Fazlollahi, R.R. Aliev
Soft Computing and its Applications in Business and Economics, 2004
ISBN 3-540-22138-7

Vol. 158. K.K. Dompere
Cost-Benefit Analysis and the Theory of Fuzzy Decisions – Identification and Measurement Theory, 2004
ISBN 3-540-22154-9

Vol. 159. E. Damiani, L.C. Jain, M. Madravia
Soft Computing in Software Engineering, 2004
ISBN 3-540-22030-5

Vol. 160. K.K. Dompere
Cost-Benefit Analysis and the Theory of Fuzzy Decisions – Fuzzy Value Theory, 2004
ISBN 3-540-22161-1

Nadia Nedjah
Luiza de Macedo Mourelle (Eds.)

Evolvable Machines

Theory & Practice

Springer

Nadia Nedjah
Luiza de Macedo Mourelle
Universidade do Estado do Rio de Janeiro
Departamento de Engenharia de Sistemas e Computação
Rua São Francisco Xavier, 524
Maracanã, CEP, 20550-900
Rio de Janeiro, RJ
Brazil
E-mail: nadia@eng.uerj.br
ldmm@eng.uerj.br

ISSN 1434-9922
ISBN 3-540-22905-1 Springer Berlin Heidelberg New York

Library of Congress Control Number: 2004110948

Springer is a part of Springer Science+Business Media
springeronline.com

Printed in Germany

Typesetting: data delivered by editors
Cover design: E. Kirchner, Springer-Verlag, Heidelberg
Printed on acid free paper 62/3020/M - 5 4 3 2 1 0

To the memory of my father Ali and my beloved mother Fatiha,

Nadia

To my beloved parents Neuza and Luiz,

Luiza

Preface

Evolutionary algorithms are computer-based solving systems, which use the evolutionary computational models as key element in their design and implementation. A variety of evolutionary algorithms have been proposed. The most popular ones are genetic algorithms. They have a conceptual base of simulating the evolution of individual structures via the Darwinian natural selection process. The process depends on the adherence of the individual structures as defined by its environment to the problem pre-determined constraints. *Genetic algorithms* are well suited to provide an efficient solution of hard problems.

Methods for artificial evolution of active components, such as programs and hardware, are rapidly developing branches of adaptive computation and adaptive engineering. The evolutionary process can produce, as results, computational expressions, e.g. algorithms, or machines, e.g. mechanical or electronic devices. The evolved components generally present creativity as well as inventiveness. Furthermore, they are usually efficient in terms of the specified requirements.

This book is devoted to reporting innovative and significant progress in automatic and evolutionary methodology of applied to machine design. Theoretical as well as practical chapters are contemplated.

The content of this book is divided into three main parts. The first part consists of four chapters while the second and third part have three chapters. In the following, we give a brief description of the main contribution of each of these chapters.

Part I: Evolvable Robots

In Chapter 1, which is entitled *Learning for Cooperative Transportation by Autonomous Humanoid Robots*, the authors, **Yutaka Inoue, Takahiro To-**

hge and Hitoshi Iba, first clarify the practical difficulties we face from the cooperative transportation task with two bodies of humanoid robots. Afterwards, we propose a solution to these difficulties and empirically show the effectiveness both by simulation and by real robots.

In Chapter 2, which is entitled *Evolution, Robustness and Adaptation of Sidewinding Locomotion of Simulated Snake-like Robot*, the authors, namely **Ivan Tanev, Thomas Ray and Andrzej Buller**, inspired by the efficient method of locomotion of the rattlesnake, propose an automatic design through genetic programming, of the fastest possible sidewinding locomotion of simulated limbless, wheelless snake-like robot or *snakebot*. this work can be considered as a step forward towards building real Snakebots that are able to perform robustly in difficult environment.

In Chapter 3, which entitled *Evolution of Khepera Robotic Controllers with Hierarchical Genetic Programming Techniques*, the authors, namely **Marcin L. Pilat and Franz Oppacher**, show how to evolve robotic controllers for a miniature mobile Khepera robot. They concentrate on control tasks for obstacle avoidance, wall following, and light avoidance. The robotic controllers are evolved through canonical GP implementation, linear genome GP system, and hierarchical GP methods (Automatically Defined Functions, Module Acquisition, Adaptive Representation through Learning).

In Chapter 4, which entitled *Evolving Controllers for Miniature Robots*, the author, namely **Michael Botros**, presents a series of experiments in evolutionary robotics that used the miniature mobile robot Khepera. Khepera robot is widely used in evolutionary experiments due to its small size and light weight which simplify the setup of the environments needed for the experiments. The controllers evolved in the presented experiments include classical and spiking neural networks controllers, fuzzy logic controllers and computer program obtained by Genetic Programming.

Part II: Evolvable Hardware Synthesis

In Chapter 5, which is entitled *Evolutionary Synthesis of Synchronous Finite State Machines*, the authors, namely **Nadia Nedjah and Luiza de Macedo Mourelle**, propose an evolutionary methodology synthesise finite state machines. First, they optimally solve the state assignment NP-complete problem, which is inherent to designing any synchronous finite state machines using genetic algorithms. This is motivated by the fact that with an optimal state assignment one can physically implement the state machine in question using a minimal hardware area and response time. Second, with the optimal state assignment provided, we propose to use the evolutionary methodology to yield optimal evolvable hardware that implement the state machine control component. The evolved hardware requires a minimal hardware area and introduces a minimal propagation delay of the machine output signals.

In Chapter 6, which is entitled *Automating the Hierarchical Synthesis of MEMS Using Evolutionary Approaches*, the authors, namely **Zhun Fan, Jiachuan Wang, Kisung Seo, Jianjun Hu, Ronald Rosenberg, Janis Terpenny and Erik Goodman**, first discuss the hierarchy that is involved in a typical MEMS design. Then they move on to discuss how evolutionary approaches can be used to automate the hierarchical design and synthesis process for MEMS. At the system level, genetic programming, as a strong search tool, is used to generate and search in the topologically open-ended design space. A multiple-resonator microsystem design is taken as an example to illustrate the integrated design automation idea using evolutionary approaches at multiple levels.

In Chapter 7, which is entitled *An Evolutionary Approach to Multi-FPGAs System Synthesis*, the authors, namely **F. Fernández de Veja, J.I. Hidalgo, J.M. Sánchez and J. Lanchares**, explain in details a methodology for Multi-FPGA systems design. They describe a set of techniques based on evolutionary algorithms, and we show that they are capable of solving all of the design tasks, which are partitioning, placement and routing. Firstly a hybrid compact genetic algorithm is used solves the partitioning problem and then genetic programming is exploited to evolve a solution for the two remaining tasks.

Part III: Evolvable Designs

In Chapter 8, which is entitled *Evolutionary Computation and Parallel Processing Applied to the Design of Multilayer Perceptrons*, the authors namely, **Ana Claudia M. L. Albuquerque, Jorge D. Melo and Adrião D. Dória Neto**, present the use of genetic algorithms in defining the neural network's architecture and in refining its synaptic weights. A different approach of a cooperative parallel genetic algorithm with different evolution behaviors is given. Applications on approximation of functions will be illustrated.

In Chapter 9, which is entitled *Evolvable Fuzzy Hardware for Real-time Embedded Control in Packet Switching*, the authors, namely **Ju Hui Li, Meng Hiot Lim, Qi Cao**, describe a scheme to implement an Evolvable Fuzzy Hardware for real-time Packet Switching Problem. The proposed evolvable fuzzy hardware addresses many issues effectively. For the hardware implementation of the evolvable fuzzy hardware, real-time fuzzy inference with high-speed context switching capability is necessary. This aspect is addressed through implementation based on a context independent reconfigurable fuzzy inference chip.

In Chapter 10, which is entitled *Improving Multi Expression Programming: An Ascending Trail from Sea-Level Even-3-Parity Problem to Alpine Even-18-Parity Problem*, the author, namely **Mihai Oltean**, proposes and uses several techniques for improving the search performed by Multi Expression Programming. Some of the most important improvements are Automatically Defined

Functions and Sub-Symbolic node representation. Several experiments with Multi Expression Programming are performed in this chapter. Numerical results show that Multi Expression Programming performs very well for the considered test problems.

Nadia Nedjah, Ph.D.
Luiza de Macedo Mourelle, Ph.D.

Department of System Engineering & Computation
Faculty of Engineering
State University of Rio de Janeiro
(nadia | ldmm)@eng.uerj.br
http://www.eng.uerj.br

Contents

Part I Evolvable Robots

Part II Evolvable Hardware Synthesis

List of Figures

List of Tables

List of Algorithms

Part I

Evolvable Robots

1

Learning for Cooperative Transportation by Autonomous Humanoid Robots

Yutaka Inoue, Takahiro Tohge, and Hitoshi Iba

Department of Frontier Informatics, Graduate School of Frontier Sciences,
The University of Tokyo, 7-3-1 Hongo, Bunkyo-ku, Tokyo, 113-8656, Japan,
inoue@iba.k.u-tokyo.ac.jp, http://www.iba.k.u-tokyo.ac.jp

In this chapter, we describe a cooperative transportation to a target position with two humanoid robots and introduce a machine learning approach to solving the problem. The difficulty of the task lies on the fact that each position shifts with the other's while they are moving. Therefore, it is necessary to correct the position in a real-time manner. However, it is difficult to generate such an action in consideration of the physical formula. We empirically show how successful the humanoid robot HOAP-1's cooperate with each other for the sake of the transportation as a result of Q-learning. Furthermore, we show a result of the experiment that transports an object cooperatively to a target position using those robots.

1.1 Introduction

In this chapter, we first clarify the practical difficulties we face from the cooperative transportation task with two bodies of humanoid robots. Afterwards, we propose a solution to these difficulties and empirically show the effectiveness both by simulation and by real robots.

In recent years, many researches have been conducted upon various aspects of humanoid robots [1] [2]. Since humanoid robots have physical features similar to us, it is very important to let them behave intelligently like humans. In addition, from the viewpoint of AI or DAI (Distributed AI), it is rewarding to study how cooperatively humanoid robots perform a task just as we humans can. However, there have been very few studies on the cooperative behaviors of multiple humanoid robots. Thus, in this chapter, we describe the emergence of the cooperation between humanoid robots so as to achieve the same goal. The target task we have chosen is a cooperative transportation, in which two bodies of humanoids have to cooperate with each other to carry and transport an object to a certain goal position.

As for the transportation task, several researches have been reported on the cooperation between a human and a wheel robot [3][4] and the cooperation among multiple wheel robots [5][6]. However, in most of these studies, the goal was to let a robot perform a task instead of a human.

Research to realize collaboration with a legged robot includes lifting operations of an object with two robots [7] and box-pushing with two robots [8]. However, few studies have addressed cooperative work using similar legged robots. It is presumed that body swinging during walking renders cooperative work by a legged robot difficult [9]. Therefore, it is more difficult for a humanoid robot to carry out a transportation task, because it is capable of motions that are more complicated and less stable than a usual legged robot.

In leader-follower type control [10][11], which is often used for cooperative movement, it is essential that a follower robot acquire information such as the position and velocity of an object fluctuated by the motion of a leader robot. This information is usually obtained by a force sensor or wireless communication. Such a method is considered to be effective for a robot with a stable center of gravity operating with less information for control. However, much information must be processed simultaneously to allow a humanoid robot to perform complicated actions, such as transporting an object cooperatively, with its difficulty to control caused by its unstable body balance. It would be expensive to build a system that carries out optimal operation using this information.

One hurdle in the case where multiple humanoid robots move carrying an object cooperatively is the disorder of cooperative motion by body swinging during walking. Therefore in this chapter, learning is carried out to acquire behavior to correct a mutual position shift generated by this disorder of motion. For this purpose, we use two kinds of methods: (i) Classifier System [12] and (ii) Q-learning [13]. We will show that behavior to correct a position shift can be acquired based on the simulation results of this study. Moreover, according to this result, the applicability to a real robot is investigated. Furthermore, cooperative transportation to a target position is conducted.

This chapter is organized as follows. The next section explains the clarified problem difficulties with the cooperative transportation. After that, Section 1.3 proposes our method to solve the problem. Section 1.4 presents an experimental result in the simulation and real robots environment. Then Section 1.5 shows an experimental result of cooperative transportation with real robots. Section 1.6 discusses these results and future researches. Finally, a conclusion is given in Section 1.7.

1.2 Problem in cooperative Transportation by humanoid Robots

Cooperative transportation by humanoid robots involves solving many difficult problems. It is different from the transportation by a single robot, in which another robot motion is negligible. On the other hand, in case of the cooperative transportation, one robot's motion has an influence on another robot to some extent. Thus, it is necessary to synchronize both robots' motions. However, the synchronization is not easily achieved because precise motions are not expected by humanoids due to the load weight or the floor friction.

We conducted an experiment assuming tasks to transport a lightweight object all around, aiming to extract specific problems from using two humanoid robots: HOAP-1 (manufactured by Fujitsu Automation Limited). Dimensions of a HOAP-1 are 223 x 139 x 483 mm (width, depth, and height) with a weight of 5.9 kg. It has two arm joints with 4 degrees of freedom each, and two leg joints with 6 degrees of freedom each: 20 degrees of freedom in all for right and left.

Actually, when a package is transferred, it seems to be more practical for two robots to have a single object. However, unless both robots move synchronously in the desirable direction, too much load will be given to the arms of robots, which may often cause the mechanical trouble in the arm and the shoulder. It is assumed in experiment that the arm movement can cancel the position shift, and that the distance and angle that can be cancelled would be in the space between two objects.

We assume the following task situation (see Fig. 1.1a): Each robot raises its platform, on which a brick, i.e., a transportation target, is to be placed. However, as a first step, we have removed the target for the sake of simplicity (Fig. 1.1b). The platform each robot raises is made of foam polystyrene and about 80 gram weigh. The size is about 150 mm wide, 150 mm deep and 200 mm high. This platform is larger than a conventional one because it has to bear the weight of the transportation target. A sponge grip is attached on each robot arm, so that an object would not slip off the arm during the experiment.

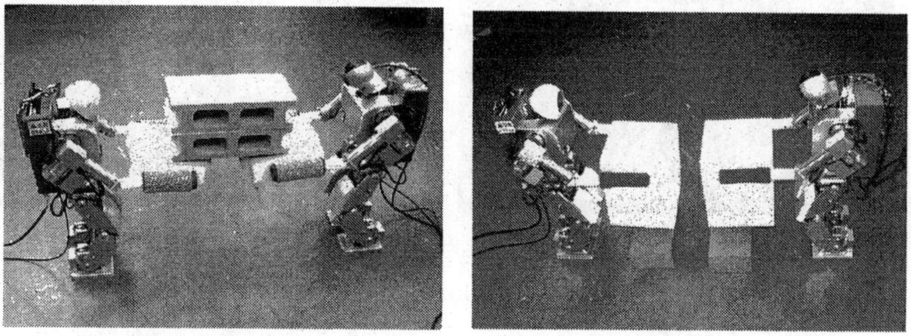

(a) Trunk-based transfer (b) Simplified transportation.

Fig. 1.1. The target of cooperative transportation.

The two robots operate in Master-Slave mode. That is, the Master robot transmits data corresponding to each operation created in advance to the Slave robot; the two robots start a motion in the same direction simultaneously. The created basic motions consist of the following 12 patterns: forward, backward, rightward, leftward, half forward, half backward, half rightward, half leftward, right turn, left turn, pick up, and put down. These basic motions are combined to allow the two robots to transport an object.

The experiment of several times was conducted using each motion. The initial position in this experiment is shown in Fig. 1.2a. The results indicated that unintentional motions such as lateral movement (Fig. 1.2b) and back-and-forth movement (Fig. 1.2c) by sliding from the normal position, and rotation (Fig. 1.2d) occur frequently in basic transportation motions such as forward, backward, rightward, and leftward. This is considered mainly to result from swinging during walking and the weight of the object.

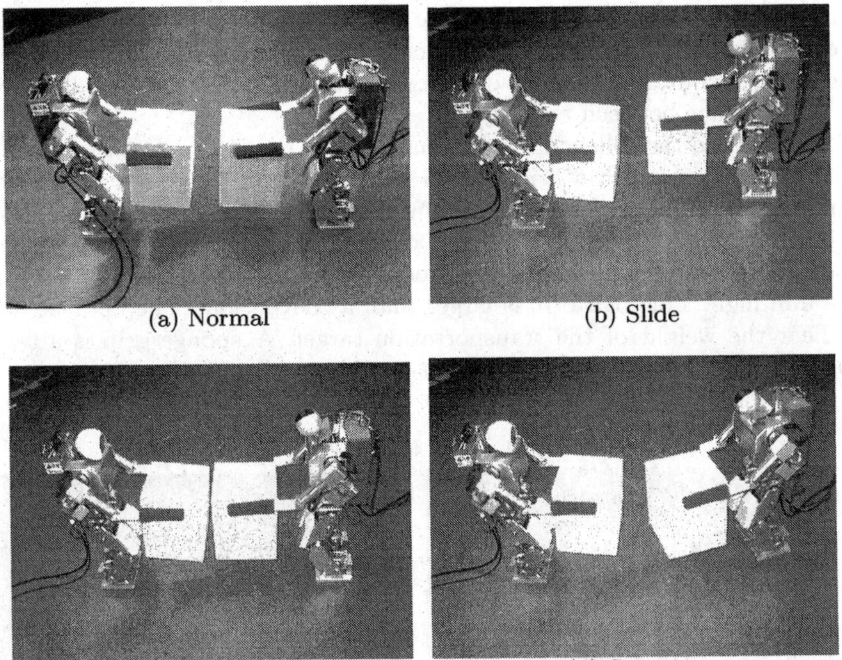

(a) Normal (b) Slide

(c) Approach (d) Spinning

Fig. 1.2. Normal positions and different kinds of positional shifts.

The following three factors can be considered the causes of these shifts in motion.

- Swing when the robot moves

- Shift of the center of gravity by having an object
- Initialization error of robot's joint motors

Especially, in case of humanoid robots, we can think of motor vibration due to the body motion as its cause. This may affect the robot's translation or direction. In addition, the gravity change resultant from carrying an object may possibly cause some errors in the movement.

When activating a robot, it is necessary to set the initial positions of each joint's motors manually. Thus, setting those ini-tial values wrongly may result in fatal errors. In order to investigate the error of initial setting, we performed experiments in the fundamental mode of motions: forward, backward, rightward and leftward. More precisely, a robot was forced to make five steps in each direction so as to measure the final position. In these experiments, the initial setting was used twice in each of test patterns, and the experiments were repeated 10 time, which means that 20 trials were performed in to-tal for each setting. Note that the same robot was used for these experiments.

The moving distance to front and back, right and left of each experiment is shown in Fig. 1.3. The moving distance in two initial setups is expressed by a circle and a triangle. As shown in Fig. 1.3a and Fig. 1.3b, when the robot moves front-ward or backward, the error occurs to the right incline. On the other hand, Fig. 1.3c and Fig. 1.3d show that rightward or leftward movements resulted in the errors in the frontward incline. From the results, it is evident that coincident ini-tial positioning of two robots is very difficult, and error occurs in moving distance or in direction.

Such a position shift can be cancelled, if only slight, by installing a force sensor on a wrist and moving arms in the load direction. However, a robot's position must be corrected in case of a shift beyond the limitation of an arm. Improper correction may cause failure of an arm or a shoulder joint and breakage of an object.

1.3 Approach of Transportation Control

The practical problem of transporting an object is the possibility that a robot falls during movement, due to loss of body balance in connection with a load on the arm by a mutual position shift after moving. Therefore, it is important to acquire behavior for correcting the position shift generated from movement by learning algorithms.

One of the advantages of using reinforcement learning is its easiness of revising the system due to the change of input-output information and its possibility to select an appropriate action in response to various information.

A situation is assumed in which two robots move face to face while maintaining the distance within a range to transport an object stably. This motion can be divided into two stages: one in which the two robots move simultaneously, and one in which one robot corrects its position. Simultaneous movement of two robots is controlled by wireless communication. A shift over a

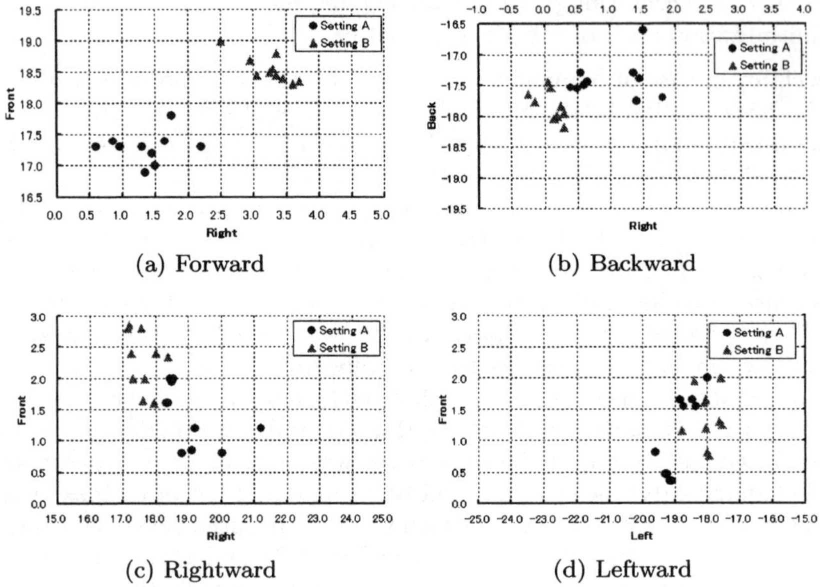

(a) Forward (b) Backward

(c) Rightward (d) Leftward

Fig. 1.3. Experimental results of initial setting.

certain limit of distance or angle in this motion will be corrected by one robot according to behavior acquired by learning.

In order to recognize an object or a state, the Master robot is equipped with an active camera, while the Slave robot carries a static one. The active camera works with a pan angle of ±90[deg]Cand a tilt angle of ±90[deg]. The robots rotate these cameras and recognize their goal so that they can transport the target object to the goal. The static camera is used to observe the current state of two robots. The obtained information is used as the input to the learning system.

Fig. 1.4 shows the motion overview for conducting a transportation task. In the first stage, the Master robot performs a motion programmed in advance; simultaneously, it issues directions to perform the same motion to the Slave robot. If there is no position shift after movement, the process forwards to the next stage; otherwise, the position is corrected with the learning system. We have tried to realize a cooperative transportation task by repeating the series of this flow.

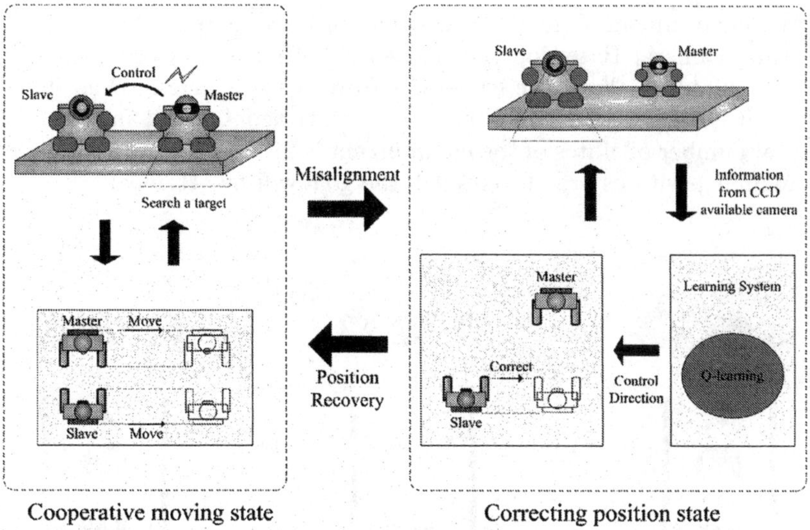

Fig. 1.4. Steps of the cooperative transportation.

1.4 Learning to Correct Positioning

1.4.1 Learning Model

The learning for position correction is carried out with Q-learning and Classifier System.

Q-learning guarantees that the state transition in the environment of a Markov decision process converges into the optimal direction [14]. However, it requires much time until the optimal behavior obtains a reward in the early stage of learning. Thus, it takes time for the convergence of learning. Furthermore, because all combinations of a state and behavior are evaluated for a predetermined Q value, it is difficult to follow environmental change. Therefore, learning by a real robot is extremely difficult because of the processing time.

On the other hand, Classifier System can learn a novel classification and to maintain the diversity by means of GA, which evolves a rule including # (don't care symbol). Thus, it enables the learning with relatively few trials so that the evolved robot may adapt the dynamic environment more effectively. However, too much generalization might result in the poor performance due to the overfitting.

We use these above two methods for the sake of simulation-based learning of the position correction and compare the obtained results.

The effective division of states and the selection of actions are very essential for the sake of efficient Q-learning and Classifier System. A static camera is attached to one robot to obtain information required for learning from the external environment. The external situation is evaluated with images from this static camera. Based on the partner robot's position projected on the image acquired by the static camera, a state space is arranged as shown in Fig. 1.5. It is divided into three states: vertical, horizontal, and angular. Hence, the total number of states of the environment is 27. If the vertical, horizontal, and angular positions are all centered, the goal will be attained.

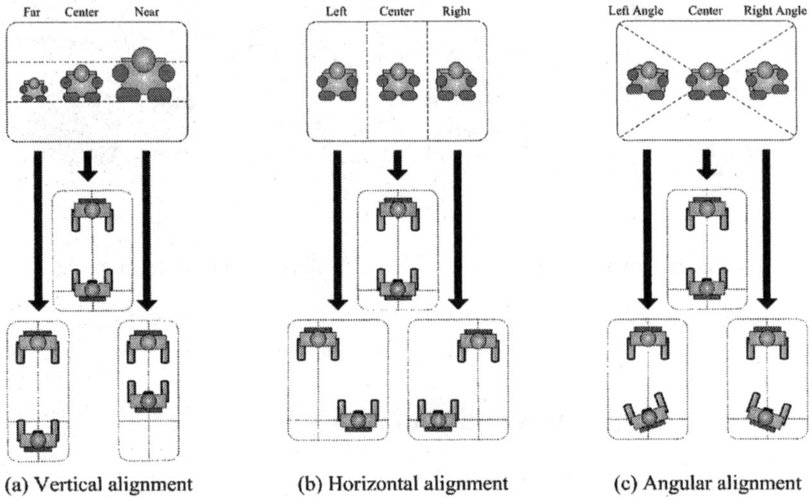

Fig. 1.5. Different states (27-states).

We assumed six behaviors which a robot can choose among the 12 patterns mentioned in Section 1.2. They are the especially important motions of forward, backward, rightward, leftward, right turn, and left turn. Fig. 1.6 depicts all these motions.

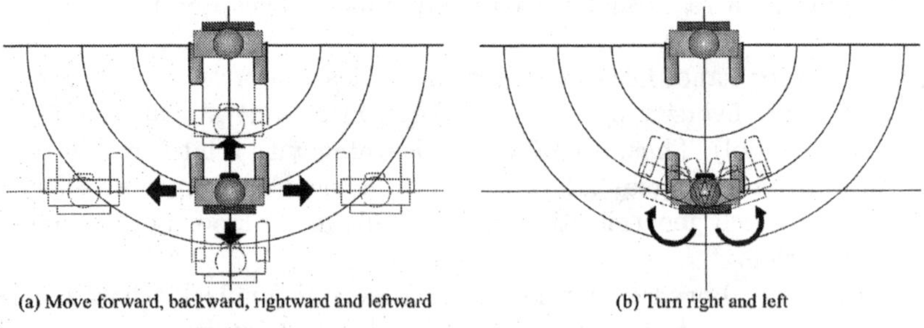

Fig. 1.6. Different actions (6-actions).

1.4.2 Learning in Simulator

The learning model stated in the preceding subsection has been realized in a simulation environment. This simulator sets a target position at a place of constant distance from the front of the partner robot, which is present in a plane. A task will be completed if the learning robot reaches the position and faces the partner robot.

The target position here ranges in distance movable in one motion. In this experiment, back-and-forth and lateral distances and the rotational angle movable in one motion are assumed to be constant. That is, if the movable distance in one step is about 10 cm back-and-forth and 5 cm laterally, the range of the target point will be 50 cm^2. In this range, the goal will be attained if the learning robot is in place where it can face the partner robot with one rotation motion.

The Q-learning parameters for the simulation were as follows: the initial Q value, Q_0, was 0.0, the learning rate α was 0.01, the reduction ratio γ was 0.8 and the reward was 1.0 for the task achievement. We used the following parameters for Classifier Systems and GA: the initial value for a rule is 0.1, the tax is 0.001, the bid value is 0.01, the crossover rate is 0.95, the mutation ratio is 0.05, and the population size is 1,024.

A certain noise is added to the motion. This is to establish the learning scheme in consideration of uncertain factors, such as translation errors due to the motion or different operational characteristics of robots. More precisely, 5% error is given to a motion at one time as noise.

1.4.3 Result of Simulator Learning

Behavior patterns obtained by simulation with the Q-learning approach in the early stage and acquired by learning are shown in Figs. 1.7a and 1.7b, respectively. In the early stage, motions are observed such as walking to the same place repeatedly and going to a direction different from the target position. Behavior approaching the target position is gradually observed as learning progresses; finally, behavior is acquired to move to the target position and turn to the front with relatively few motions.

As can be seen Classifier System simulation by in Figs. 1.7c and 1.7d, the trajectory divergence occurred at the earlier stage of learning. However, at the later generations, the effective actions were acquired so as to face the goal correctly.

Fig. 1.8a plots the success rate of learning for 1,000 steps. Fig. 1.8b gives the number of successful motions with generations. Both data were averaged over 10 runs. As can be seen, Q-learning is superior. This may be because it enables hill-climbing local search. Classifier System's performance goes up and down irregularly. However, this is considered to show the superiority in terms of the robust learning. As a result of this, numbers of motions are almost the same for both methods as the later stage of learning.

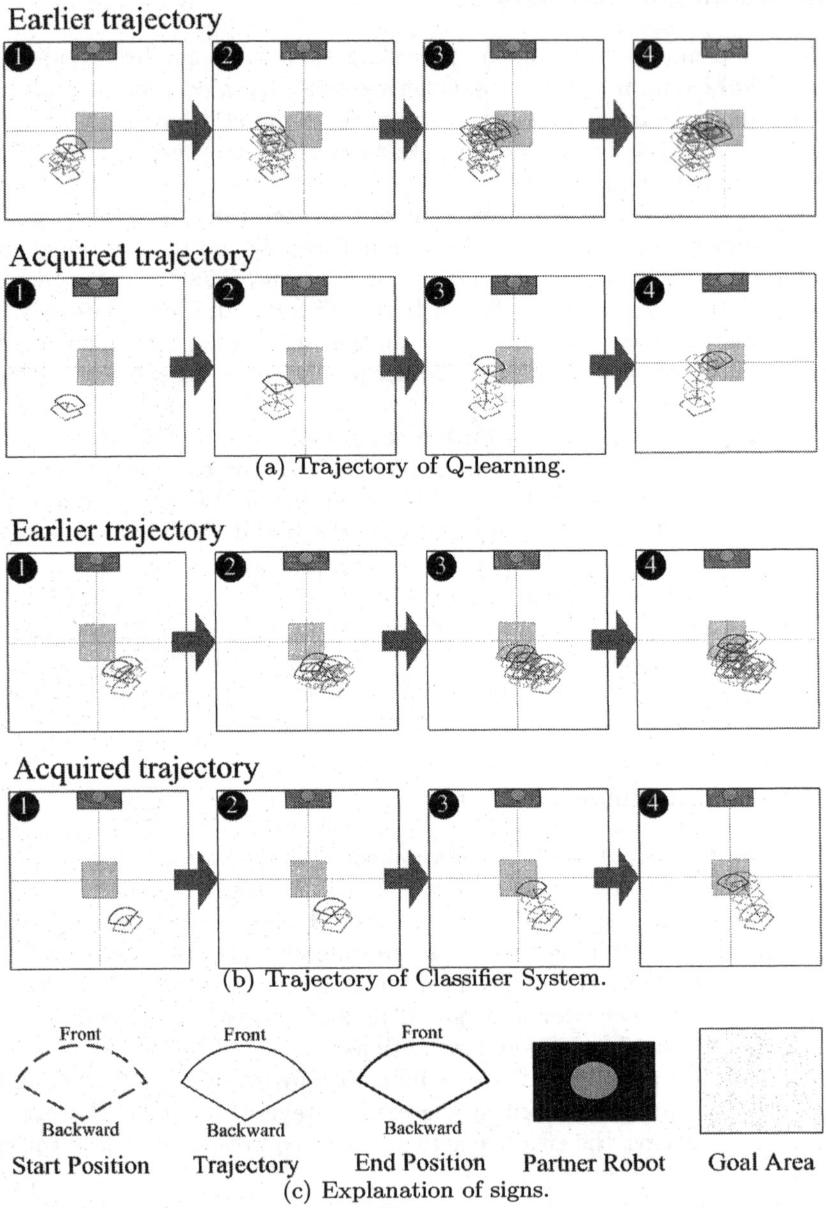

Fig. 1.7. Results of a simulation with Q-learning and Classifier System.

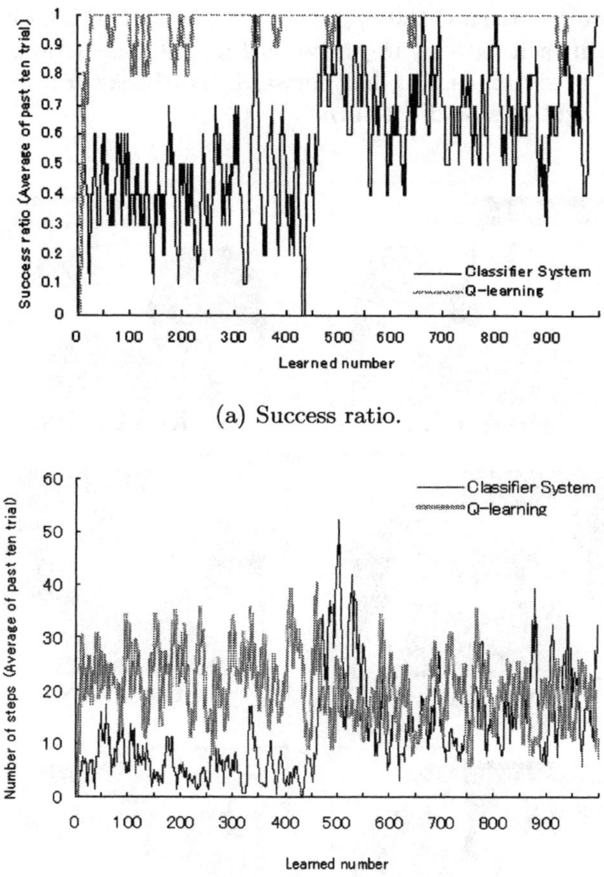

(a) Success ratio.

(b) Moved number.

Fig. 1.8. Q-learning vs. Classifier System.

1.4.4 Experiments with Real Robots

Following the simulation results described in the previous subsection, we conducted an experiment with real robots to con-firm their applicability. In this experiment, we have used the learning data obtained from Q-learning, because Q-learning acquired the relatively more precise behaviors than Classifier System in the previous simulation.

For the recovery from the horizontal left (right) slide, a humanoid robot was initially shifted leftward (rightward) against the opponent robot by 5.2 cm. On the other hand, it was initially moved forward (backward) from the correct position by 3.2 cm for the recovery from front (back) position. In case of the rotation failure, the robot was shifted either leftward or rightward by

5.2 cm and rotated toward the opponent by 20 degrees. The images of the static camera in each pattern are shown in Fig. 1.9. The actions used for the recovery were of six kinds, i.e., half forward, half backward, half rightward, half leftward, right turn and left turn.

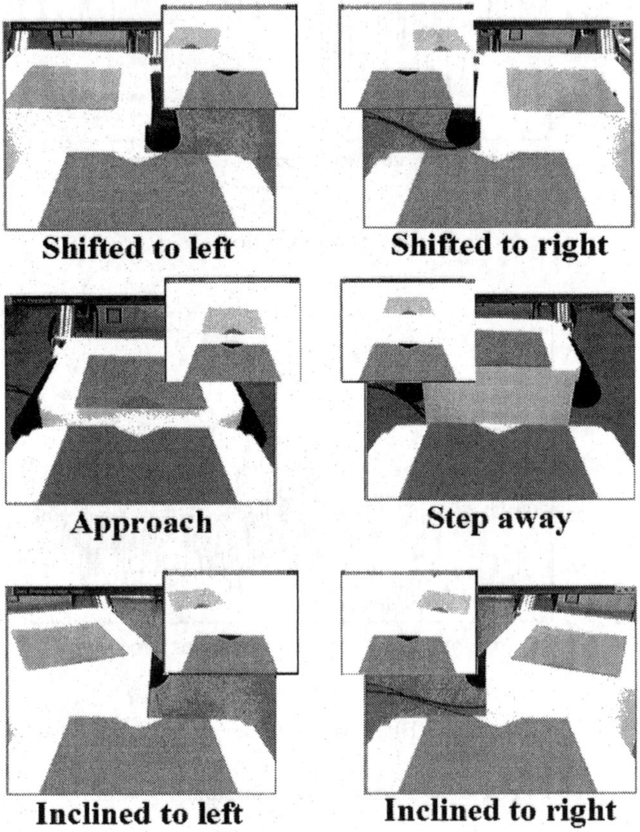

Fig. 1.9. Type of the experiments.

For this experiment, robots started from one of the three patterns shown in Figs. 1.2b, 1.2c and 1.2d, which were classified as the failure of actions (see Section 1.2). We employed two HOAP-1's, one of which used the learning results, i.e., the acquired Q-table, so as to generate actions for the sake of recovery from the failure. Q-learning was conducted by simulation with different numbers of iterations, i.e., 1,000, 10,000, and 100,000 iterations. The learning parameters were the same as in the previous subsection.

1.4.5 Experimental results

Table 1.1 shows the averaged numbers of actions for the sake of recovery from the above three failure patterns. In Table 1.1: RL represents the slide recovery from the right, LR is the slide recovery from the left, NF stands for the distance recovery from the front, FN is defined as the distance recovery from the back, RLS and LRS are respectively the angle recovery from the right and from the left. The averaged numbers of required actions were measured over five runs for each experimental condition, i.e., with different Q-learning iterations.

Table 1.1. Numbers of average movement.

		Q-learning Iterations		
Failure	Recovery	1,000times	10,000times	100,000times
Horizontal slide	RL	4.6	4.4	4.4
	LR	5.4	5.2	5.2
Approach and away	NF	6.6	1.8	1.8
	FN	2.0	2.0	1.4
Spinning around	RLS	9.4	9.4	8.6
	LRS	11.4	16.8	10.2

For slide motion, the robot learned an effective motion after 1,000 time steps. This is explained in the following way. A gap usually occurs even when a robot corrects a position. However, correcting a slide position requires only a simple sequence of actions, as a result of which the gap rarely occurs.

With 1,000 iterations, more actions were needed to recover from the front position to the back. This is because the robot had acquired the wrong habit of moving leftward when the opponent robot was approaching (see Fig. 1.10). This habit has been corrected with 10,000 iterations, so that much fewer actions were required for the purpose of repositioning.

The recovery from "spinning around" seems to be the most difficult among the three patterns. For this task, the movement from the slant to the front (see Fig. 1.11) was observed with 10,000 iterations, which resulted in the increase of required actions. This action sequence was not observed with 1,000 iterations. This is considered that the phenomenon is caused by the difference between simulation and a real-world environment.

1.5 Cooperative Transportation to Target Position

1.5.1 Experiments with Real robots

The cooperative transportation task, i.e., two humanoid robots cooperate with each other to transport an object to a certain goal, is carried out by using the

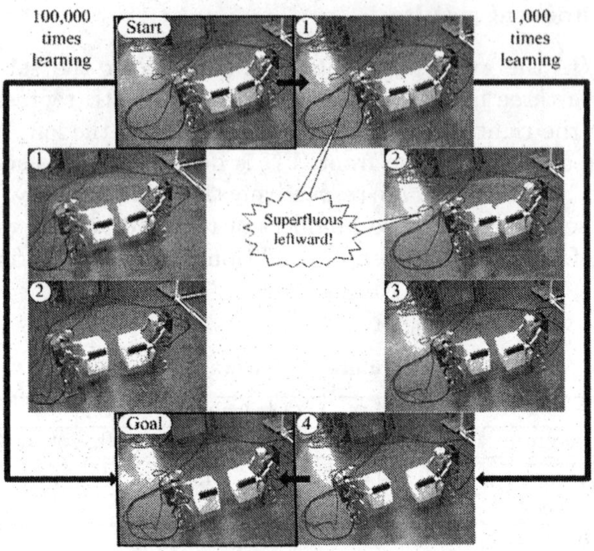

Fig. 1.10. Behavior of NF with short-time learning and full learning.

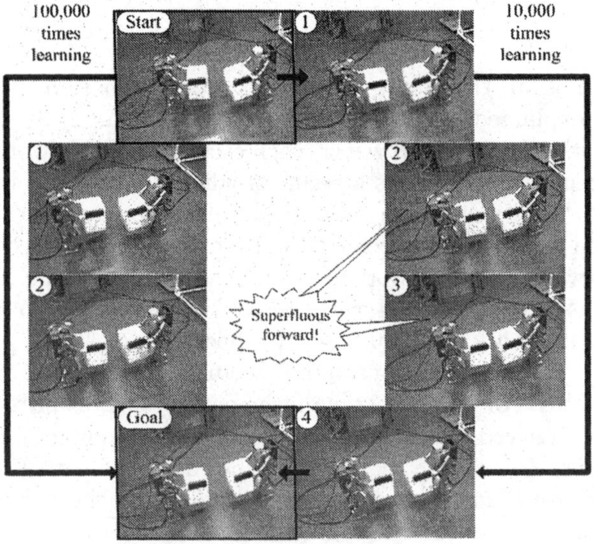

Fig. 1.11. Behavior of LRS with short-time learning and full learning.

obtained Q-learning data shown in the previous section. The transportation target is a sphere made of foam polystyrene. Its diameter is about 25 cm and 63 gram weigh. The goal is positioned in a place about 1m distant from each humanoid robot and is marked for the purpose of recognition.

The Master robot finds its mark using the active camera, and decides the transportation path to the destination. The path is derived as follows:

1. Move the Master robot forward or backward so that it is next to the goal.
2. Move the Master robot left or right to a position adjacent to the mark.

In the meantime, if a positional shift occurs, the Slave robot recognizes its type and tries to recover from it. Afterward, the Master robot searches for a new path again and the transportation is restarted according to the new path.

1.5.2 Experimental results

Fig. 1.12 shows the transportation process with some recovery actions. As can be seen, two recovery actions were performed in case of side motions. As a result, the robots achieved the task successfully. In case of a position shift, the path to the goal was slightly changed. This was caused by each other's shift and its recovery. In order to reduce this anomaly and re-calculation of the path, two robots need to revise their positions simultaneously.

Moreover, when the goal is seen overlapped with the opponent robot, the mark is difficult to recognize. In order to solve this difficulty, two robots should rotate cooperatively with the object on the platform or both robots should be equipped with active cameras for the recognition.

1.6 Discussion

We have established a learning system for the cooperative transportation by simulation and confirmed its real-world applicability by means of real robots. Furthermore, we have conducted cooperative transportation including acquired behavior to correct position using real robots.

The effective actions were acquired for the sake of recovery from the position failure as a result of simulation learning. In a real environment, at the earlier stage of learning, we have often observed the unexpected movement to a wrong direction by real humanoid robots; which was also the case with the simulation. In the middle of learning, the forward movement was more often observed from the slant direction. These types of movements, in fact, had resulted in the better learning performance by simulation, whereas in a real environment they prevented the robot from moving effectively. This is considered to be the distinction between simulation and a real-world environment. We have confirmed the success of the cooperative transportation by real robots, i.e., both robots cooperatively transported an object to a goal while revising their position shift effectively.

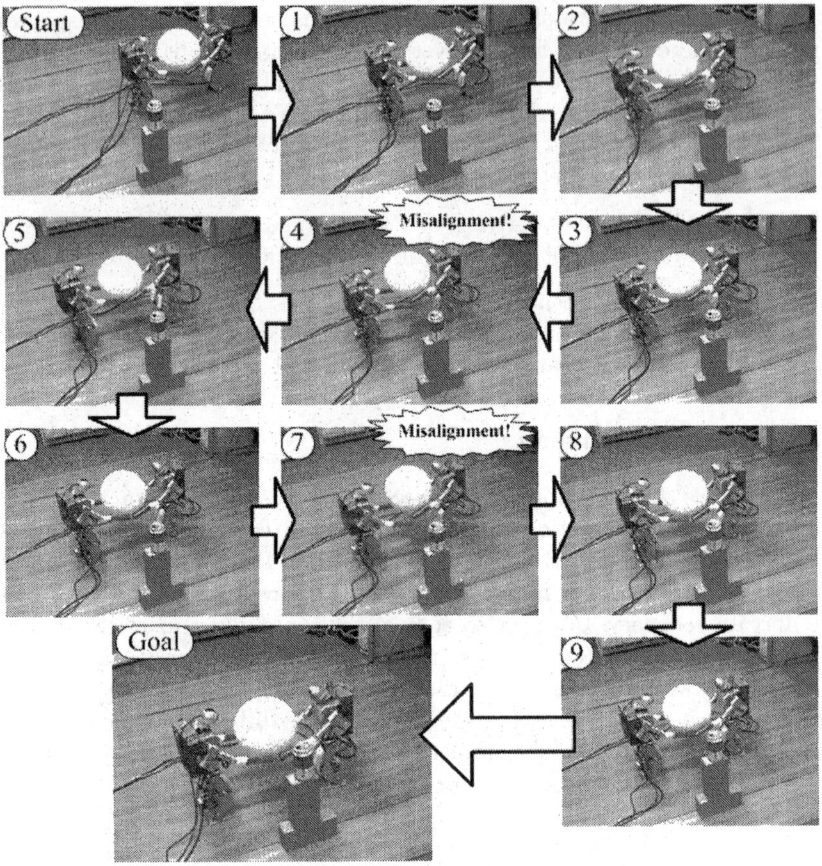

Fig. 1.12. Result of an experiment with real robots.

In this chapter, the position recovery was carried out by one robot. It is more desirable and efficient if both robots can do so. For this purpose, the learning of two robots in a real environment is essential. This is also important to nullify the difference between simulation and real-world environment. However, it is not easy using Q-learning because of the frequent loss of a goal or an opponent in the early state of the learning in a real environment. Thus, we can conclude Classifier System is superior to Q-learning for the purpose of the cooperative learning in a real-world environment.

Moreover, we are now developing a methodology of filtering learning result by means of camera information from difference devices, for the purpose of applying the obtained result in a simulator to a real environment. This method is based on the evolutionary computation and probabilistic estimation.

In order to solve the difficulty with the distinction, learning in the real world is essential. For this purpose, we are currently working on the integration of GP and Q-learning in a real robot environment [15]. This method does not need a precise simulator, because it is learned with a real robot. In other words, the precision requirement is met only if the task is expressed properly. As a result of this idea, we can greatly reduce the cost to make the simulator highly precise and acquire the optimal program by which a real robot can perform well. We especially showed the effectiveness of this approach with various types of real robots, e.g. SONY AIBO or HOAP-1.

1.7 Conclusion

Specific problems were extracted in an experiment using a practical system in an attempt to transport an object cooperatively with two humanoid robots. The result proved that both body swinging during movement and the shift in the center of gravity, by transporting an object, caused a shift in the position after movement.

We investigated the behavior of fundamental motions to make sure the impact of initial positioning on the robot operation. Consequently, it is found that position matching of motors is very difficult even using the same robot and even in the same motion, there occur errors in moving distance and direction.

Therefore, we have proposed a learning method to revise a position shift while the cooperative transportation, and established a learning framework in a simulation. In addition, the obtained results were verified by using real robots in a real environment.

In order to move towards the target position efficiently, it is necessary to perform the real learning by two robots. Therefore, it is important to discuss the approach for efficient movement and perform experiment with real robots. Since huge time is required for learning in real robots, it is important to reduce the time of learning in real environment using learning data in the simulator.

In our future work, we want to study how robots can more to the target in the shortest path when there is an obstacle in the path or how to more in an L-shaped path.

References

1. K. Yokoi, et al., "Humanoid Robot's Application in HRP", In *Proc. of IARP International Workshop on Humanoid and Human Friendly Robotics*, pp.134-141, 2002.
2. H. Inoue, et al., "Humanoid Robotics Project of MITI", *The 1st IEEE-RAS International Conference on Humanoid Robots*, Boston, 2000.
3. O. M. AI-Jarrah and Y. F. Zheng, "Armmanipulator Coordination for Load Sharing using Variable Compliance Control", In *Proc. of the 1997 IEEE International Conference on Robotics and Automation*, pp.895-900, 1997.

4. M. M. Rahman, R. Ikeura and K. Mizutani, "Investigating the Impedance Characteristics of Human Arm for Development of Robots to Cooperate with Human Operators", In *CD-ROM of the 1999 IEEE International Conference on Systems, Man and Cybernetics*, pp.676-681, 1999.
5. N. Miyata, J. Ota, Y. Aiyama, J.Sasaki and T. Arai, "Cooperative Transport System with Regrasping Car-like Mobile Robots", In *Proc. of the 1997 IEEE/RSJ International Conference on Intelligent Robots and Systems*, pp.1754-1761, 1997.
6. H. Osumi, H.Nojiri, Y.Kuribayashi and T.Okazaki, "Cooperative Control of Three Mobile Robots for Transporting A Large Object", In *Proc. of International Conference on Machine Automation (ICMA2000)*, pp.421-426, 2000.
7. M. J. Matarić, M. Nillson and K. T. Simsarian, "Cooperative Multi-robot Box-pushing", In *Proc. of the 1995 IEEE/RSJ International Conference on Intelligent Robots and Systems*, pp.556-561, 1995.
8. H.Kimura and G.Kajiura, "Motion Recognition Based Cooperation between Human Operating Robot and Autonomous Assistant Robot", In *Proc. of the 1997 IEEE International Conference on Robotics and Automation*, pp.297-302, 1997.
9. Y. Inoue, T. Tohge and H. Iba, "Cooperative Transportation by Humanoid Robots - Learning to Correct Positioning -", In *Proc. of the Hybrid Intelligent Systems (HIS2003)*, pp.1124-1133, 2003.
10. J. Ota, Y. Buei, T. Arai, H. Osumi and K. Suyama, "Transferring Control by Cooperation of Two Mobile Robots", *The Journal of the Robotics Society of Japan (JRSJ)*, vol.14, no.2, pp.263-270, 1996.
11. K. Kosuge, T. Osumi and H. Seki, "Decentralized Control of Multiple Manipulators Handling an Object", In *Proc. of the 1996 IEEE/RSJ International Conference on Intelligent Robots and Systems*, pp.318-323, 1996.
12. L. B. Booker, D. E. Goldberg, and J. H. Holland, "Classifier Systems and Genetic Algorithms", In *Machine Learning: Paradigms and Methods*, *MIT Press*, 1990.
13. R. S. Sutton and A. G. Barto, "Reinforcement Learning", *MIT Press*, Boston, 1998.
14. C. J. C. H. Watkins and P. Dayan, "Q-learning", *Machine Learning*, Vol. 8, pp.279-292, 1992.
15. S. Kamio, H. Mitsuhashi and H. Iba, "Integration of Genetic Programming and Reinforcement Learning for Real Robots", In *Proc. of the Genetic Computation Conference (GECCO2003)*, pp.470-477, 2003.

2

Evolution, Robustness and Adaptation of Sidewinding Locomotion of Simulated Snake-like Robot

Ivan Tanev[1], Thomas Ray[2], and Andrzej Buller[3]

[1] Department of Information Systems Design,
Faculty of Engineering, Doshisha University, Japan
ATR Network Informatics Laboratories, Keihanna Science City, Kyoto, Japan
i_tanev@atr.jp

[2] Department of Zoology, University of Oklahoma Norman, Oklahoma, USA
ATR Human Information Science Laboratories,
Keihanna Science City, Kyoto, Japan
ray@atr.jp

[3] ATR Network Informatics Laboratories Keihanna Science City, Kyoto, Japan
buller@atr.jp

Inspired by the efficient method of locomotion of the rattlesnake *Crotalus cerastes*, the objective of this work is automatic design through genetic programming, of the fastest possible (sidewinding) locomotion of simulated limbless, wheelless snake-like robot (Snakebot). The realism of simulation is ensured by employing the Open Dynamics Engine (ODE), which facilitates implementation of all physical forces, resulting from the actuators, joints constrains, frictions, gravity, and collisions. Empirically obtained results demonstrate the emergence of sidewinding locomotion from relatively simple motion patterns of morphological segments. Robustness of the sidewinding Snakebot, considered as ability to retain its velocity when situated in unanticipated environment, is illustrated by the ease with which Snakebot overcomes various types of obstacles such as a pile of or burial under boxes, rugged terrain and small walls. The ability of Snakebot to adapt to partial damage by gradually improving its velocity characteristics is discussed. Discovering compensatory locomotion traits, Snakebot recovers completely from single damage and recovers a major extent of its original velocity when more significant damage is inflicted. Contributing to the better understanding of sidewinding locomotion, this work could be considered as a step towards building real Snakebots, which are able to perform robustly in difficult environment.

2.1 Introduction

Wheelless, limbless snake-like robots (Snakebots) feature potential robustness characteristics beyond the capabilities of most wheeled and legged vehicles - ability to traverse terrain that would pose problems for traditional wheeled or legged robots, and insignificant performance degradation when partial damage is inflicted. Moreover, due to their modular design, Snakebots may be cheaper to build, maintain and repair. Some useful features of Snakebots include smaller size of the cross-sectional areas, stability, ability to operate in difficult terrain, good traction, high redundancy, and complete sealing of the internal mechanisms [3], [4]. Robots with these properties open up several critical applications in exploration, reconnaissance, medicine and inspection. However, compared to the wheeled and legged vehicles, Snakebots feature (i) smaller payload, (ii) more difficult thermal control, (iii) more difficult control of locomotion gaits and (iv) inferior speed characteristics. Considering the first two drawbacks as beyond the scope of our work, and focusing on the drawbacks of control and speed, we intend to address the following challenge: how to develop control sequences of Snakebot's actuators, which allow for achieving the fastest possible speed of locomotion.

Although for many tasks, handcrafting the robot locomotion control code by applying various theoretical approaches [1],[2], [13], [15], [19], [21] can be seen as a natural approach, it might not be feasible for developing the control code of Snakebot due to its morphological complexity. While the overall locomotion gait of Snakebot might emerge from relatively simply defined motion patterns of morphological segments of Snakebot, neither the degree of optimality of the developed code nor the way to incrementally improve the code is evident to the human designer [11]. Thus, an automated mechanism for solution evaluation and corresponding rules for incremental optimization of the intermediate solution(s) are needed [5], [10]. The proposed approach of employing genetic programming (GP) implies that the code, which governs the locomotion of Snakebot is automatically designed by a computer system via simulated evolution through selection and survival of the fittest in a way similar to the evolution of species in the nature. The use of an automated process to design the control code opens the possibility of creating a solution that would be better than one designed by a human [9].

Out choice of employing GP for designing the control code of snakebot is motivated by the commonly accepted recognition that applying traditional learning techniques (reinforcement learning, Q-learning, etc.) to redundant robotic systems (such as snake-like robots) is difficult and extremely time consuming. These learning techniques usually involve many random searches, and the number of these random searches increases exponentially with the increase of size of the search space.

Evolving a Snakebot's locomotion (and in general, behavior of any robot) could be performed as a first step in the sequence of simulated off-line evolution (phylogenetic learning) on the software model, followed by on-line adaptation (ontogenetic learning) of evolved code on a physical robot situated in a real environment [12]. Off-line software simulation facilitates the process of Snakebot's controller design because the verification of behavior on physical Snakebot is extremely time consuming, costly and often dangerous for Snakebot and surrounding environment. Moreover, in some cases it is appropriate to initially model not only the locomotion, but also to co-evolve the most appropriate morphology of the artifact (i.e. number of phenotypic segments; types and parameters of joints which link segments; actuators' power; type, amount and location of sensors; etc.) [14], [16], [17] and only then (if appropriate) to physically implement it as hardware. The software model, used to simulate Snakebot should fulfill the basic requirements of being quickly developed, adequate, and fast running [6]. Typically slow development time of GP stems from the highly specific semantics of the main attributes of GP (e.g. representation, genetic operations, fitness evaluation) and can be significantly reduced through incorporating off-the-shelf software components and open standards in software engineering. To address this issue, we developed a GP framework based on open XML standard and ensure adequacy and runtime efficiency of Snakebot simulation, we applied the Open Dynamic Engine (ODE) freeware software library for simulation of rigid body dynamics.

The *objectives* of our work are (i) to explore the feasibility of applying GP for automatic design of the fastest possible locomotion of realistically simulated Snakebot and (ii) to investigate the robustness and adaptation of such locomotion to unanticipated environmental conditions and degraded abilities of Snakebot. Inspired by the fast sidewinding locomotion of the rattlesnake *Crotalus cerastes*, this work is motivated by our desires (i) to better understand the mechanisms underlying sidewinding locomotion of natural snakes, (ii) to explore the phenomenon of emergence of locomotion of complex bodies from simply defined motion patterns of the morphological segments comprising these bodies, (iii) to verify the feasibility of employing ODE for realistic software simulation of a Snakebot, and (iv) to investigate the practicality of building real Snakebots.

The remainder of this document is organized as follows. Section 2.2 emphasizes the main features of the GP proposed for evolution of locomotion of simulated Snakebot. Section 2.2 presents empirical results of evolving locomotion gaits of Snakebots and discusses the emergence of sidewinding. The same section elaborates on robustness and adaptation of sidewinding to unanticipated environmental conditions and partial damage of Snakebot. Finally, Section 2.4 draws a conclusion.

2.2 Approach

2.2.1 Representation of Snakebot

Snakebot is simulated as a set of identical spherical morphological segments ("vertebrae"), linked together via universal joints. All joints feature identical (finite) angle limits and each joint has two attached actuators ("muscles"). In the initial, standstill position of Snakebot the rotation axes of the actuators are oriented vertically (vertical actuator) and horizontally (horizontal actuator) and perform rotation of the joint in the horizontal and vertical planes respectively (Fig. 2.1). Considering the representation of Snakebot, the task of designing the fastest locomotion can be rephrased as developing temporal patterns of desired turning angles of horizontal and vertical actuators of each segment, that result in fastest overall locomotion of Snakebot.

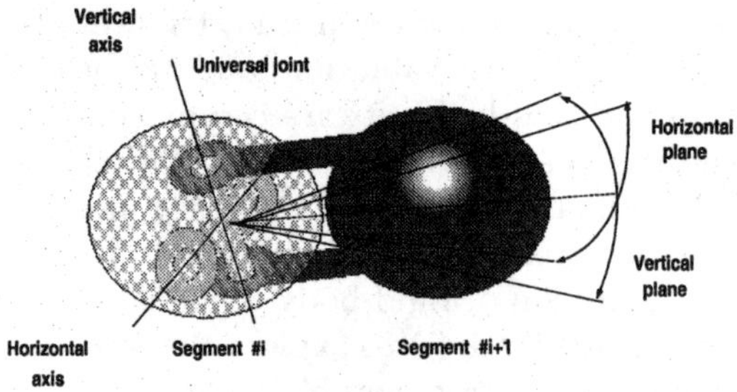

Fig. 2.1. Morphological segments of Snakebot linked via universal joint. Horizontal and vertical actuators attached to the joint perform rotation of the segment #i+1 in vertical and horizontal planes respectively

2.2.2 Algorithmic paradigm

Genetic Programming

GP [7] is a domain-independent problem-solving approach in which a population of computer programs (individuals' genotypes) is evolved to solve problems. The simulated evolution in GP is based on the Darwinian principle of reproduction and survival of the fittest. The fitness of each individual is based on the quality with which the phenotype of the simulated individual is performing in a given environment. The major attributes of GP - function set, terminal set, fitness evaluation, genetic representation, and genetic operations are elaborated in the remaining of this Section.

Function set and terminal Set

In applying GP to evolution of Snakebot, the genotype is associated with two algebraic expressions, which represent the temporal patterns of desired turning angles of both the horizontal and vertical actuators of each morphological segment. Since locomotion gaits are periodical, we include the trigonometric functions sin and cos in the GP function set in addition to the basic algebraic functions. The choice of these trigonometric functions reflects our intention to verify the hypothesis (first expressed by Petr Miturich in 1920's) that undulative motion mechanisms could yield efficient gaits of snake-like artifacts operating in air, land, or water. Terminal symbols include the variables time, index of morphological segment of Snakebot, and two constants: Pi, and random constant within the range [0, 2]. The main parameters of the GP are summarised in Table 2.1. The rationale of employing automatically defined function (ADF) is based on empirical observation that the evolvability of straightforward, independent encoding of desired turning angles of both horizontal and vertical actuators is poor, although it allows GP to adequately explore the search space and ultimately, to discover the areas which correspond to fast locomotion gaits in solution space. We discovered that (i) the motion patterns of horizontal and vertical actuators of each segment in fast locomotion gaits are highly correlated (e.g. by frequency, direction, etc.) and that (ii) discovering and preserving such correlation by GP is associated with enormous computational effort. ADF, as a way of introducing modularity and reuse of code in GP [8] is employed in our approach to allow GP to explicitly evolve the correlation between motion patterns of horizontal and vertical actuators as shared fragments in algebraic expressions of desired turning angles of actuators. Moreover, the best result was obtained by (i) allowing the use of ADF as a terminal symbol in algebraic expression of desired turning angle of vertical actuator only, and (ii) by evaluating the value of ADF by equalizing it to the value of currently evaluated algebraic expression of desired turning angle of horizontal actuator.

Table 2.1. Main parameters of GP

Category	Value
Function set	sin, cos, +, -, *, /
Terminal set	time, segment_ID, Pi, random constant, ADF
Population size	200 individuals
Selection	Binary tournament, ratio 0.1
Elitism	Best 4 individuals
Mutation	Random subtree mutation, ratio 0.01
Fitness	Velocity of simulated Snakebot during the trial
Trial interval	180 time steps, each time step account for 50ms of real time
Termination criterion	(Fitness1 $\geq$ 100) or (Generations $\geq$ 30) or no improvement of fitness for 16 generations)

Fitness evaluation

The fitness function is based on the velocity of Snakebot, estimated from the distance which the center of the mass of Snakebot travels during the trial. The real values of the raw fitness, which are usually within the range (0, 2) are multiplied by a normalizing coefficient in order to deal with integer fitness values within the range (0, 200). A normalized fitness of 100 (one of the termination criteria shown in Table 2.1) is equivalent to a velocity which displaced Snakebot a distance equal to twice its length. The fitness evaluation routine is shown in Algorithm 2.1. The implementation of fitness evaluation routine is illustrated in 2.2.

Representation of genotype

Inspired by its flexibility, and the recently emerged widespread adoption of document object model (DOM) and extensible markup language (XML), we represent evolved genotypes of simulated Snakebot as DOM-parse trees featuring equivalent flat XML-text in a way as first implemented in [20]. Our approach implies that both (i) the calculation of the desired turning angles during fitness evaluation (functions EvalHorisontalAngle and EvalVerticalAngle, shown in Algorithm 2.2, lines 18 and 20 respectively) and (ii) the genetic operations are performed on DOM-parse trees using off-the shelf, platform- and language neutral DOM-parsers. The corresponding XML-text representation (rather than S-expression) is used as a flat file format, feasible for migration of genetic programs among the computational nodes in an eventual distributed implementation of the GP. The benefits of using DOM/XML-based representations of genetic programs are (i) fast prototyping of GP by using standard built-in API of DOM-parsers for traversing and manipulating genetic programs, (ii) generic support for the representation of grammar of strongly-typed GP using W3C-standardized XML-schema; and (iii) inherent Web-compliance of eventual parallel distributed implementation of GP.

Genetic operations

Binary tournament selection is employed - a robust, commonly used selection mechanism, which has proved to be efficient and simple to code. Crossover operation is defined in a strongly typed way in that only the DOM-nodes (and corresponding DOM-subtrees) of the same data type (i.e. labeled with the same tag) from parents can be swapped. The sub-tree mutation is allowed in strongly typed way in that a random node in genetic program is replaced by syntactically correct sub-tree. The mutation routine refers to the data type of currently altered node and applies randomly chosen rule from the set of applicable rewriting rules as defined in the grammar of strongly typed GP.

ODE

We have chosen Open Dynamics Engine (ODE) [18] to provide a realistic simulation of physics in applying forces to phenotypic segments of Snakebot, for simulation of Snakebot locomotion. ODE is a free, industrial quality software library for simulating articulated rigid body dynamics. It is fast, flexible and robust, and it has built-in collision detection. The ODE-related parameters of simulated Snakebot are summarized in Table 2.2.

Algorithm 2.1 Fitness evaluation routine

Step 1. Incorporating the evolved genotype into the actuators' controllers of Snakebot;
Step 2. Simulating the locomotion of Snakebot governed by current actuators' controllers;
Step 3. Estimating the distance, which the center of the mass of Snakebot travels during the trial.

Table 2.2. ODE-related parameters of simulated Snakebot

Parameter	Value
Number of phenotypic segments in snake	15
Model of segment	Sphere, R=0.2
Type of joint between segments	Universal
Initial alignment of segments in Snakebot	Along Y-axis of the world
Number of actuators per joint	2
Orientation of axes of actuators	Horizontal - along X-axis and Vertical - along Z-axis of the world
Operational mode of actuators	dAMotorEuler
Max force of actuators	12
Actuators stops (angular limits)	50
Friction between segments and surface (μ)	5
Sampling frequency of simulation	20 Hz

2.3 Experimental Results

This section discusses experimental results verifying the feasibility of applying GP for evolution of the fastest possible locomotion gaits of Snakebot for various fitness and environmental conditions. In addition, it investigates the properties of the fastest locomotion gait, evolved in an unconstrained environment from two perspectives: (i) robustness to various unanticipated environmental conditions and (ii) gradual adaptation to degraded mechanical

Algorithm 2.2 Implementation of fitness evaluation routine

```
function Evaluate(GenH, GenV: TGenotype): real;
// GenH and GenV is a pair of algebraic expressions, which define the
// turning angle of the horizontal and vertical actuators at the joints
// of simulated Snakebot. GenH and GenV represent the evolved genotype.
const
  TimeSteps =180; // duration of the trial
  SegmentsInSnakebot=15; // # of phenotypic segments in simulated Snakebot
var
  t, s : integer;
  AngleH, AngleV : real; // desired turning angles of actuators
  CurrAngleH, CurrAngleV: real; // current turning angles of actuators
  InitialPos, FinalPos : 3DVector; // (X,Y,Z)
begin
  InitialPos:=GetPosOfCenterOfMassOfSnakebot;
   for t:=0 to TimeSteps-1 do begin
    for s:=0 to SegmentsInSnakebot-1 do begin
    // traversing XML/DOM-based GenH using DOM-parser:
    AngleH := EvalHorizontalAngle(GenH,s,t);
    // traversing XML/DOM-based GenV using DOM-parser:
    AngleV := EvalVerticalAngle(GenV,s,t);
    CurrAngleH := GetCurrentAngleH(s);
    CurrAngleV := GetCurrentAngleV(s);
    SetDesiredVelocityH(CurrAngleH-AngleH,s);
    SetDesiredVelocityV(CurrAngleV-AngleV,s);
    end;
   // detect collisions between the objects (phenotypic segments,
   // ground plane, etc.):
   dSpaceCollide;
   // Obtain new properties (position, orientation, velocity
   // vectors, etc.) of morphological segments of Snakebot as a result
   // of applying all forces:
   dWorldStep;
   end;
  FinalPos := GetPosOfCenterOfMassOfSnakebot;
  return GetDistance(InitialPos, FinalPos)/(TimeSteps);
end;
```

abilities of Snakebot. These challenges are considered as relevant for successful accomplishment of various practical tasks during anticipated exploration, reconnaissance, medicine and inspection missions.

2.3.1 Evolution of fastest locomotion gaits

Fig. 2.2 shows the fitness convergence characteristics of 10 independent runs of GP and Fig. 2.3 shows a sample snapshots of evolved best-of-run locomotion gaits when fitness is measured in any direction in an unconstrained environment. Despite the fact that fitness is unconstrained and measured as velocity in any direction, *sidewinding* locomotion (defined as locomotion predominantly perpendicular to the long axis of Snakebot) emerged in all 10 independent runs of GP, suggesting that it provides superior speed characteristics for Snakebot morphology. The dynamic motions of the sample evolved best-of-run Snakebot is illustrated in Fig. 2.4. The normalized algebraic expressions of the genotype of sample best-of-run genetic program are shown in Equations 2.1 and 2.2. The value of the automatically defined function ADF in Equation 2.2, is evaluated by equalizing it to the value of GenH, evaluated in Equation 2.1.

$$GenH = (sin(((sin(-8))*(segment\_id-time))+(3*time)))/(sin(-8)) \quad (2.1)$$

$$GenV = sin(ADF) \quad (2.2)$$

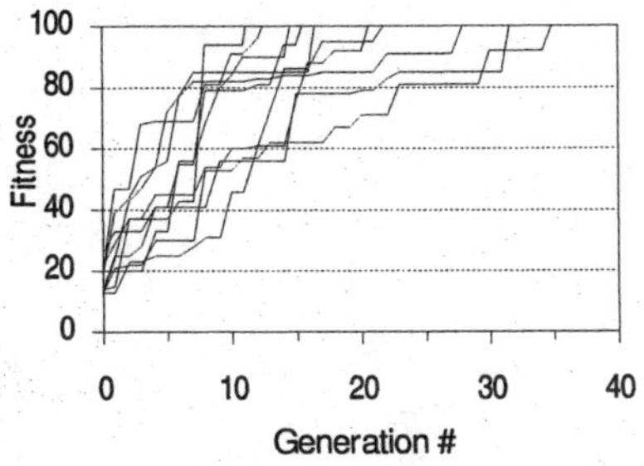

Fig. 2.2. Fitness convergence characteristics of 10 independent runs of GP for cases where fitness is measured as velocity in any direction

The dynamics of evolved turning angles of actuators in sidewinding locomotion result in characteristic circular motion pattern of segments around the center of the mass as shown in Fig. 2.5a. The circular motion pattern of segments and the characteristic track on the ground as a series of diagonal lines (Fig. 2.5b) suggest that during sidewinding the shape of Snakebot takes the form of a rolling helix. Fig. 2.5 demonstrates that the simulated evolution of locomotion via GP is able to invent the improvised cylinder of the sidewinding Snakebot to achieve fast locomotion. By modulating the oscillations of the actuators along the snake's body, the diameter of the cross-section of the "cylinder" can be tapered towards either the tail or head of the snake, providing an efficient way of "steering" the Snakebot (Fig. 2.6). Fig. 2.7 illustrates the ability of Snakebot to perform sharp turn with radius similar to its length in both clockwise and counterclockwise directions.

The moving Snakebot straight is wrapped around an imagined cylinder taking the form of a rolling helix (a). By modulating the oscillations of the

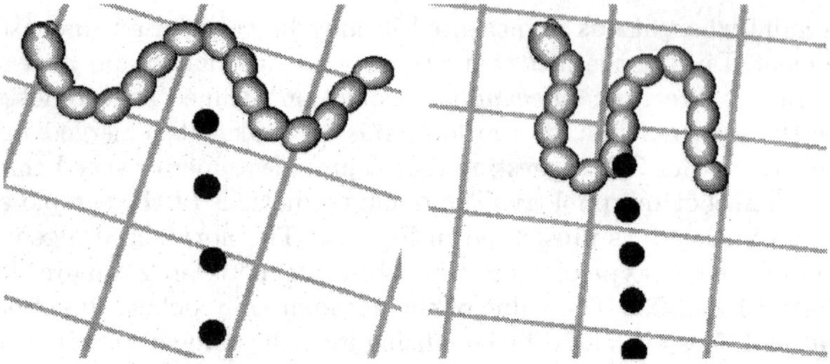

(a) Snapshots 1 (b) Snapshots 2

Fig. 2.3. Snapshots of sample evolved best-of-run sidewinding locomotion gaits of simulated Snakebot viewed from above. The dark trailing circles depict the trajectory of the center of the mass of Snakebot. Timestamp interval between each of these circles is fixed and it is the same (10 time steps) for both snapshots

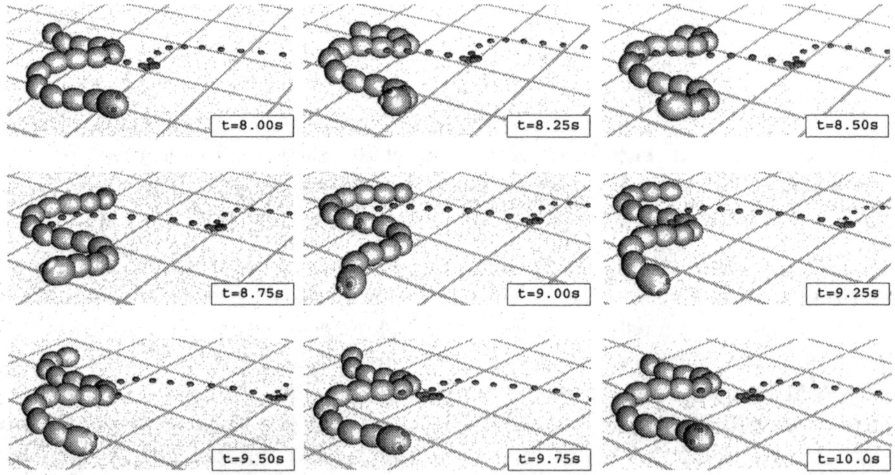

Fig. 2.4. Snapshots of sample evolved best-of-run sidewinding locomotion of simulated Snakebot (left-right top-down). The dark trailing circles depict the trajectory of the central segment of Snakebot. Timestamp interval between each of these circles is 2 time steps (0.1s)

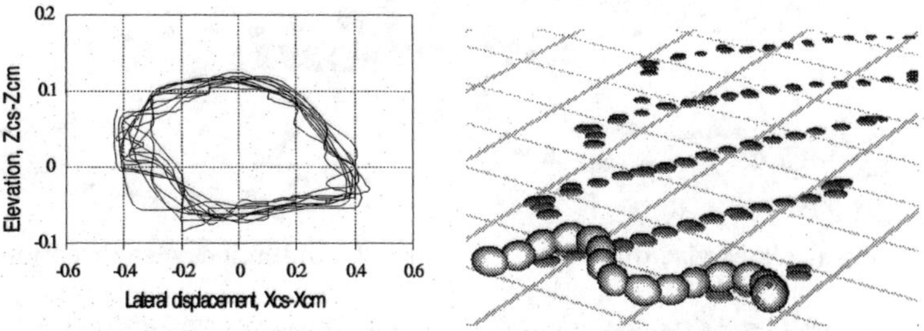

(a) Evolved best-of-run (b) Traces of ground contacts

Fig. 2.5. Trajectory of the central segment (cs) around the center of mass (cm) of Snakebot

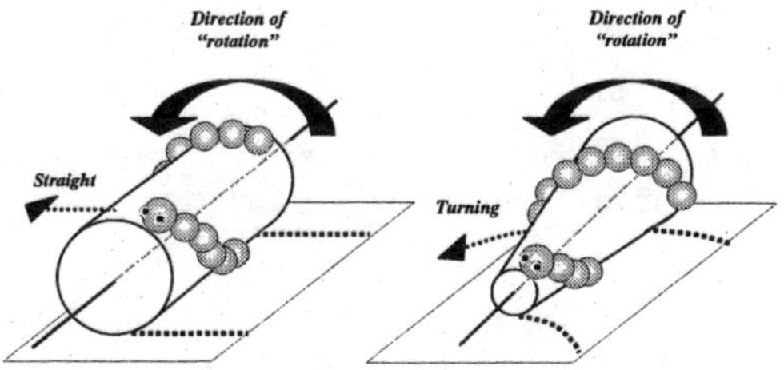

(a) Imagined cylinder (b) Imagined cone

Fig. 2.6. Steering the Snakebot

actuators along the snake's body, the diameter of the cross-section of the "cylinder" can be tapered towards either the tail or head of the snake, providing an efficient way of "steering" the Snakebot: (b) illustrates the Snakebot turning counterclockwise. The images are idealized: in simulated Snakebot (and in snakes in Nature too) the cross sectional areas of the imagined "cylinder" (a) and "cone" (b) are much more similar to ellipses (as shown in Fig. 2.5a) rather than to perfect circles as depicted in Fig. 2.6

In order to verify the superiority of velocity characteristics of sidewinding locomotion for Snakebot morphology we compared the fitness convergence characteristics of evolution in unconstrained environment for the following two cases: (i) unconstrained fitness measured as velocity in any direction (as discussed above and illustrated in Fig. 2.2 and 2.3, and (ii) fitness, measured as

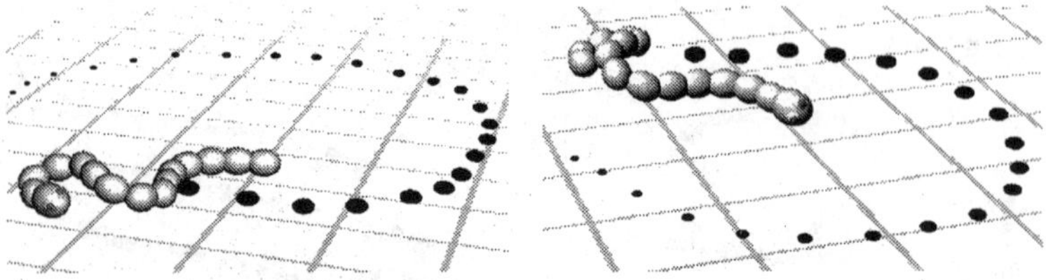

(a) Clockwise direction (b) Counterclockwise direction

Fig. 2.7. Snapshots of Snakebot performing sharp turns

velocity in forward (non-sidewinding) direction only. Fig. 2.8 depicts the fitness convergence characteristics of 10 independent runs of GP for cases where fitness is measured as velocity in forward direction. The results of evolution of forward locomotion, shown in Fig. 2.9 indicate that non-sidewinding motion, compared to sidewinding, features much inferior velocity characteristics.

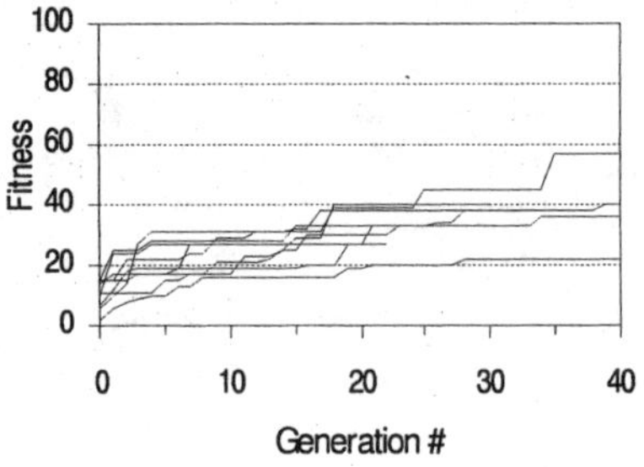

Fig. 2.8. Fitness convergence characteristics of 10 independent runs of GP for cases where fitness is measured as velocity in forward direction

The results of evolution of rectilinear locomotion of simulated Snakebot confined in narrow "tunnel" are shown in Fig. 2.10 and Fig. 2.11. The width of the tunnel is three times the diameter of the cross-section (which equals to the diameter of the segment) of Snakebot. Compared to forward locomotion in unconstrained environment (Fig. 2.8), the velocity in this experiment is superior, and comparable to the velocity of sidewinding (Fig. 2.2). This, seemingly anomalous phenomenon demonstrates the ability of simulated evolution to discover a way to utilize the walls of "tunnel" as a source of (i) extra grip and (ii) locomotion gaits which are fast yet unbalanced in an unconstrained environment. As Fig. 2.11b illustrates, as soon as Snakebot clears the tunnel, the gait flattens and velocity (visually estimated as a distance between the traces of the center of gravity of Snakebot) drops dramatically.

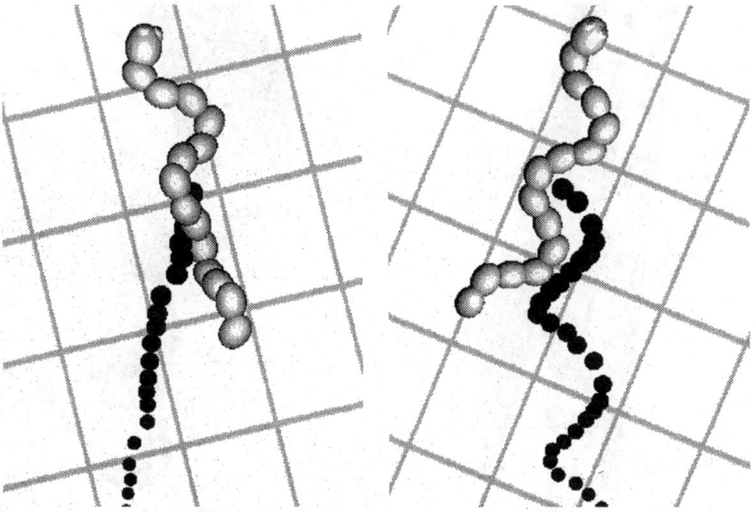

(a) Forward (b) Locomotion gaits

Fig. 2.9. Snapshots of sample evolved best-of-run forward crawling locomotion gaits of simulated Snakebot. Timestamp interval between the traces of the center of the mass is the same as for sidewinding locomotion gaits, shown in Fig. 2.3. The distance between the traces of center of the mass in both forward and sidewinding locomotion gaits comparatively illustrates the achieved velocity in both cases

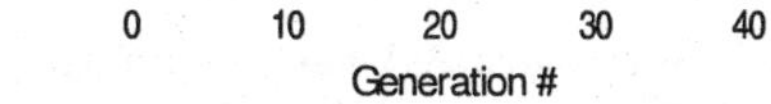

Fig. 2.10. Fitness convergence characteristics of 10 independent runs of GP when simulated Snakebot is confined in narrow "tunnel"

The final experiment discussed in this section is intended to verify the ability of GP to evolve not only periodic locomotion gaits but also standstill postures, such as elevation of the head of Snakebot. The best-of-run postures (as shown in Fig. 2.12) feature well-balanced, standstill elevation of the head. The elevation is approximately 3 diameters of Snakebot's segments, or about 20% of overall length of creature.

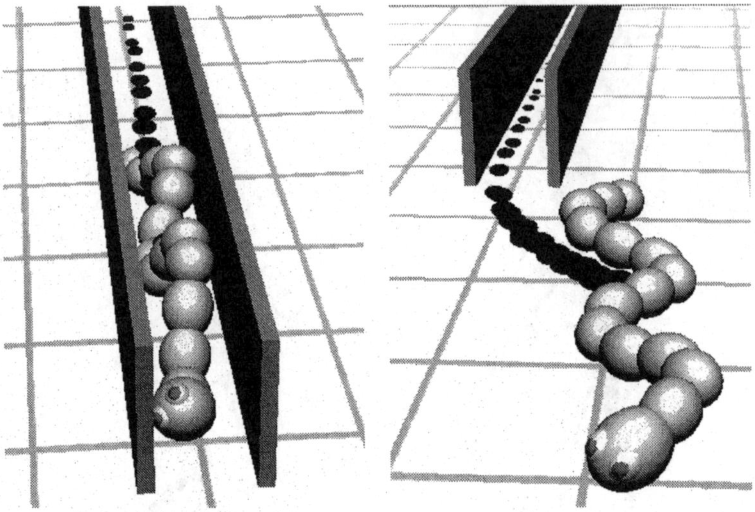

(a) Intermediate stage (b) Final stage

Fig. 2.11. Snapshots of sample evolved best-of-run gaits at the intermediate and final stages of the trial when simulated Snakebot is confined in narrow "tunnel"

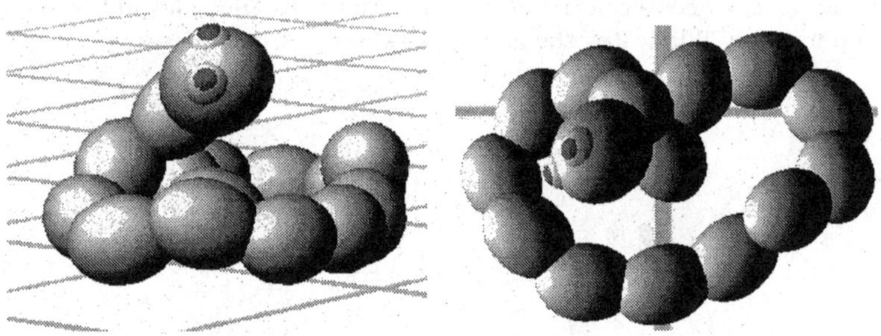

(a) Front view (b) Above view

Fig. 2.12. Snapshots of sample evolved best-of-run standstill postures featuring elevated head of Snakebot: front view and view from above

2.3.2 Robustness of Evolved Sidewinding Locomotion

Within the scope of our work we consider the robustness of sidewinding locomotion as the ability of the sidewinding Snakebot to retain its velocity when situated in a challenging environment. Robustness is qualitatively demonstrated by the ease with which the sidewinding Snakebot, initially evolved in unconstrained environment overcomes a pile of 80 boxes (Fig. 2.13), burial under 80 boxes (Fig. 2.14), rugged terrain with 200 randomly positioned obstacles with uniform random distribution of size in the range 0.1 to 1 of the diameter of the cross-section of Snakebot (Fig. 2.15) and, finally, walls with height equal to the diameter of the cross-section of Snakebot (Fig. 2.16).

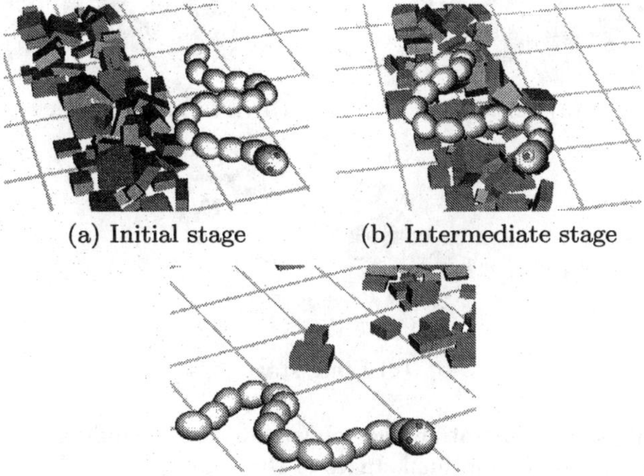

(a) Initial stage (b) Intermediate stage

(c) Final stage

Fig. 2.13. Snapshots illustrating the robustness of sidewinding in clearing a pile of boxes: initial, intermediate and final stages of the trial

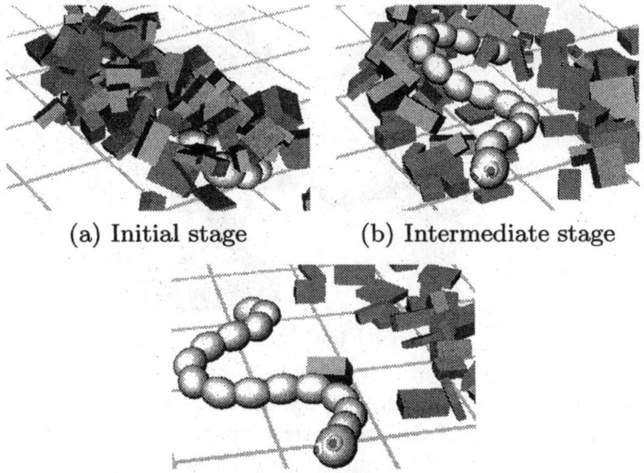

(a) Initial stage (b) Intermediate stage

(c) Final stage

Fig. 2.14. Snapshots illustrating the robustness of sidewinding in emerging from burial under a stack of boxes: initial, intermediate and final stages of the trial

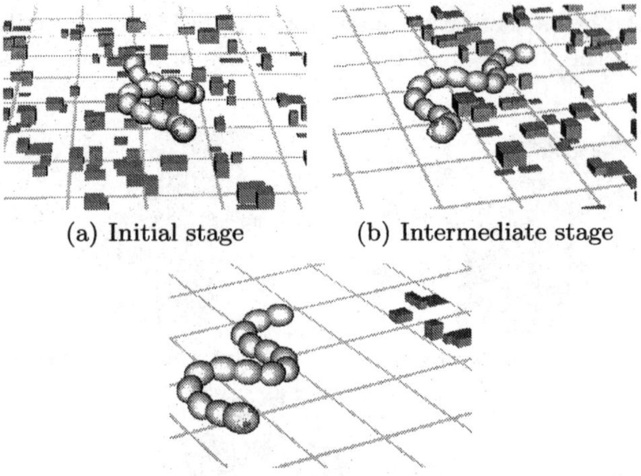

(a) Initial stage (b) Intermediate stage

(c) Final stage

Fig. 2.15. Snapshots illustrating the robustness of sidewinding in rugged terrain area: initial, intermediate and final stages of the trial

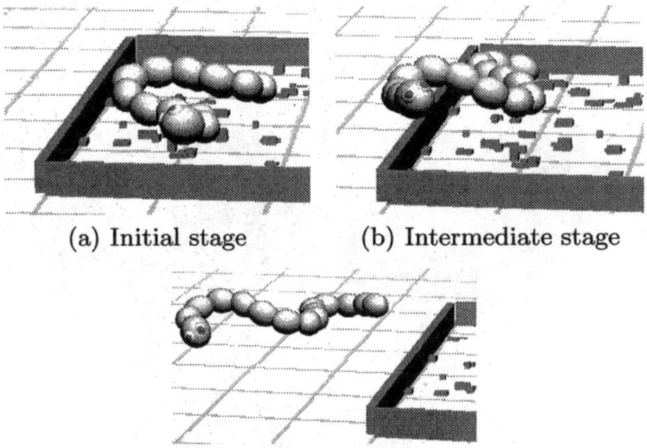

(a) Initial stage (b) Intermediate stage

(c) Final stage

Fig. 2.16. Snapshots illustrating the ability of simulated sidewinding Snakebot in clearing walls forming a "pen": initial, intermediate and final stages of the trial. Height of the walls is equal to the diameter of cross-section of simulated Snakebot

2.3.3 Adaptation

The ability of sidewinding Snakebot to adapt to partial damage to 1, 2, 4 and 8 (out of 15) segments by gradually improving its velocity by simulated evolution via GP is shown in Fig. 2.17. Demonstrated results are averaged over 4 independent runs for each case, where GP is initialized with a population comprising 190 randomly created individuals, plus 10 best-of-run genetic programs obtained from experiments with evolving sidewinding in an unconstrained environment as elaborated in Section 2.3.1. The damaged segments are evenly distributed along the body of Snakebot. Damage inflicted to a particular segment implies a complete loss of functionality of both horizontal and vertical actuators of the corresponding joint. The results of validating the adapted damaged Snakebot against the fixed best-of-run program are shown in Fig. 2.17. As Fig. 2.17 illustrates, Snakebot completely recovers from damage to single segment in 25 generations, attaining its previous velocity, and recovers to average of 94% of its previous velocity in the case where 2 (13% of total amount of 15) segments are damaged. With 4 (27%) and 8 (53%) damaged segments the degree of recovery is 77% (23% degradation) and 64% (36% degradation) respectively. Fig. 2.18a shows a snapshot of frontal view of sidewinding Snakebot adapted to damage of a single segment. Compared to the sidewinding locomotion of Snakebot before the adaptation (Fig. 2.18b), the adapted locomotion gait features much higher elevation of the middle part of the body. This elevation compensates the complete lack of functionality of actuators in the damaged segment. Snapshots of the sidewinding Snakebot are shown in Fig. 2.19, before damage to a single segment, immediately after damage to the segment, and after having completely recovered from the damage by adaptation.

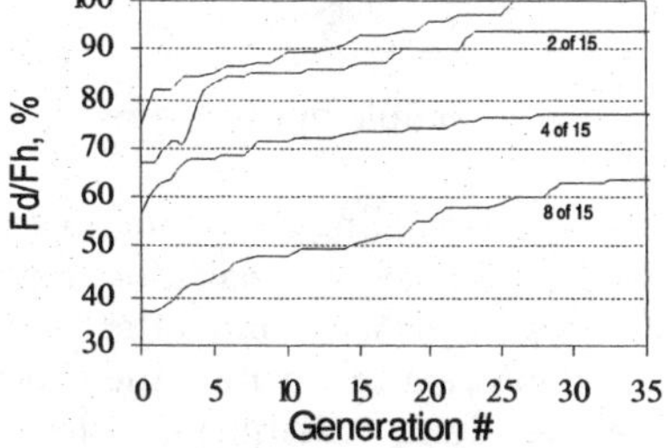

Fig. 2.17. Representation of Fd, the best fitness in evolved population of damaged snakebots, and Fh the best fitness of 10 best-of-run healthy sidewinding snakebots when sidewinding Snakebot is adapting to damage of 1, 2, 4 and 8 segments

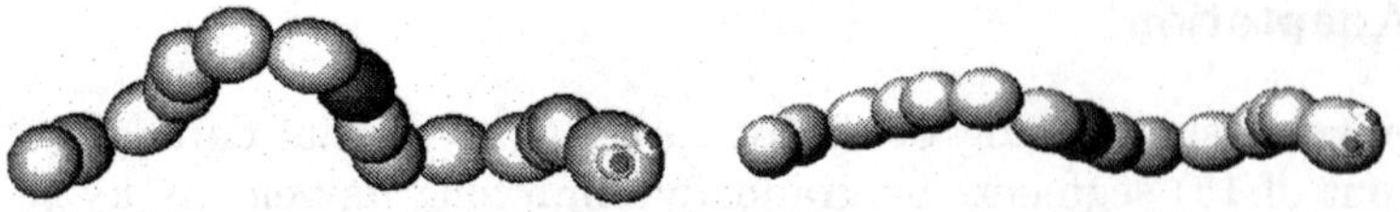

(a) After adaptation to damage (b) Before adaptation to damage

Fig. 2.18. Adaptation of sidewinding Snakebot to damage of a single segment

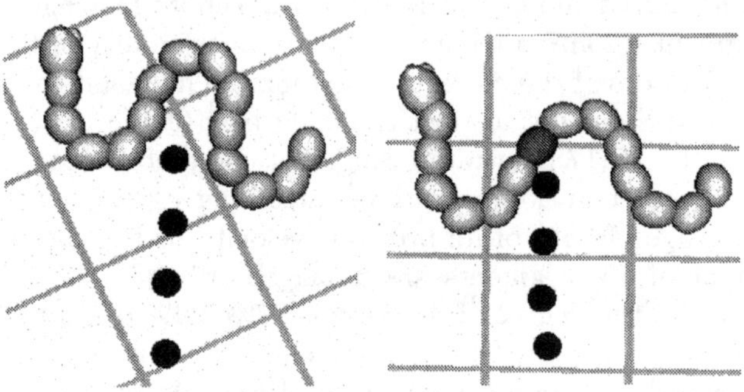

(a) Healthy Snakebot (b) After the damage

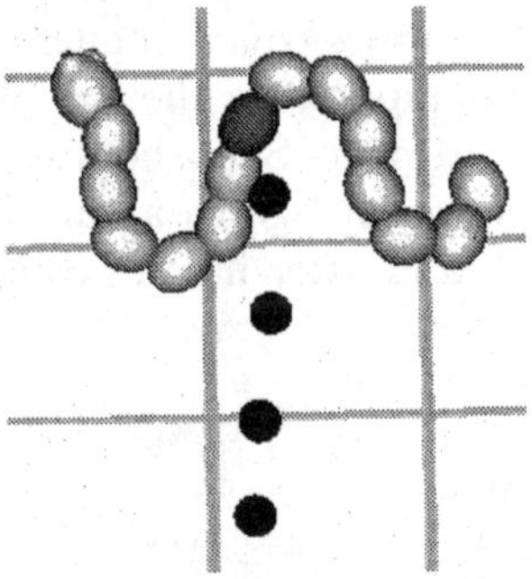

(c) Completely recovered

Fig. 2.19. Adaptation of the sidewinding Snakebot to damage of a single segment (shown in dark color): healthy Snakebot, Snakebot immediately after damage to segment #7 causing 24% loss of velocity and after having completely recovered from the damage through adaptation (c). Notice the shorter distances between the traces of the center of the mass (and consequently, slower locomotion) in case (b) compared to both (a) and (c). Snapshots (b) and (c) depict the same positions of Snakebot as shown in Fig. 2.18b and Fig. 2.18a respectively, viewed from above

2.4 Summary

We presented an approach to automatic design through genetic programming, of sidewinding locomotion of simulated limbless, wheelless artifacts. The software model used to simulate Snakebot should fulfill the basic requirements of being quickly developed, adequate, and fast running. To address the first of these issues, we employed an XML-based GP framework. To address the issues of adequacy and runtime efficiency of Snakebot simulation we applied the Open Dynamic Engine (ODE) - a freeware software library for simulation of rigid body dynamics. The empirically obtained results demonstrate that the complex locomotion of sidewinding emerges from relatively simple motion patterns of phenotypic segments (vertebrae). The evolved locomotion pattern of each segment is such that the segment is rotating in a circle-like trajectory around the center of the mass of the simulated Snakebot. This suggests that evolved sidewinding locomotion can be viewed as a process of rolling of the body of the simulated Snakebot in a helix shape, effectively inventing a kind of improvised wheel. The efficiency of sidewinding locomotion is much superior to locomotion in the forward direction, suggesting that sidewinding is the fastest possible locomotion for the simulated limbless wheelless robots with the characteristics used in this study (morphology, limits of actuator forces, joint type, joint movement limits, etc.). Robustness of the sidewinding Snakebot, initially evolved in unconstrained environment (considered as ability to retain its velocity when situated in unanticipated environment) was illustrated by the ease with which Snakebot overcomes various types of obstacles such as piles of and burial under boxes, rugged terrain and walls. The ability of Snakebot to adapt to partial damage by gradually improving its velocity characteristics was discussed. Discovering compensatory locomotion traits, Snakebot recovers completely from single damage and recovers a major extent of its original velocity when more significant damage is inflicted. Contributing to the better understanding of sidewinding locomotion, this work could be considered as a step towards building real limbless, wheelless robots, which featuring unique engineering characteristics are able to perform robustly in difficult environments.

Acknowledgements

The authors thank Katsunori Shimohara for his immense support of this research. The research was supported in part by the National Institute of Information and Communications Technology of Japan.

References

1. J. W. Burdick, J. Radford, and G.S. Chirikjian, A 'sidewinding' locomotion gait for hyper-redundant robots, In Proceedings of the IEEE int. conf. on Robotics

& Automation, Atlanta, USA, pp. 101-106, 1993
2. G. S. Chirikjian and J. W. Burdick, The kinematics of hyper-redundant robotic locomotion, IEEE Trans. Robotics and Automation, vol.11, No.6, 1995, pp. 781-793
3. K. Dowling, Limbless locomotion: Learning to Crawl with a Snake Robot, doctoral dissertation, Tech. report CMU-RI-TR-97-48, Robotics Institute, Carnegie Mellon University, 1997
4. S. Hirose, Biologically Inspired Robots: Snake-like Locomotors and Manipulators, Oxford University Press, 1993
5. S. Takamura, G. S. Hornby, T. Yamamoto, J. Yokono, and M. Fujita, Evolution of Dynamic Gaits for a Robot, IEEE International Conference on Consumer Electronics, pp. 192-193, 2000
6. N. Jacobi, Minimal Simulations for Evolutionary Robotics, Ph.D. thesis, School of Cognitive and Computing Sciences, Sussex University, 1998
7. J. R. Koza, Genetic Programming: On the Programming of Computers by Means of Natural Selection, Cambridge, MA, MIT Press, 1992
8. J. R. Koza, Genetic Programming 2: Automatic Discovery of Reusable Programs, The MIT Press, Cambridge, MA, 1994
9. J. R. Koza, M. A. Keane, J. Yu, F. H. Bennett III, W. Mydlowec, Automatic Creation of Human-Competitive Programs and Controllers by Means of Genetic Programming, Genetic Programming and Evolvable Machines, Vol.1 No.1-2, pp.121-164, 2000
10. S. Mahdavi, P. J. Bentley, Evolving Motion of Robots with Muscles, In Proc. of EvoROB2003, the 2nd European Workshop on Evolutionary Robotics, EuroGP-2003, , pp. 655-664, 2003
11. H. J. Morowitz, The Emergence of Everything: How the World Became Complex, Oxford University Press, New York, 2002
12. L. Meeden, D. Kumar, Trends in Evolutionary Robotics, Soft Computing for Intelligent Robotic Systems, edited by L.C. Jain and T. Fukuda, Physica-Verlag, New York, NY, pp. 215-233, 1998
13. J. Ostrowski and J. Burdick, Gait kinematics for a serpentine robot, In Proc. IEEE Int. Conf. on Rob. and Autom., Minneapolis, MN, pp. 1294-1299, 1996
14. R. Pfeifer, On the Role of Morphology and Materials in Adaptive Behavior, In: J.-A. Meyer, A. Berthoz, D. Floreano, H. Roitblat, and S.W. Wilson (eds.), From animals to animats 6: Proc. of the 6th Int. Conf. on Simulation of Adaptive Behavior. Cambridge, Mass., MIT Press, pp. 23-32, 2000
15. B. Salemi, P. Will, and W.-M. Shen, Distributed Task Negotiation in Modular Robots, Robotics Society of Japan, Special Issue on "Modular Robots", 2003
16. K. Sims, Evolving 3D Morphology and Behavior by Competition, Artificial Life IV Proceedings, MIT Press, pp. 28-39, 1994
17. T. Ray, Aesthetically Evolved Virtual Pets, Leonardo, Vol.34, No.4, pp. 313 - 316, 2001
18. R. Smith, Open Dynamics Engine (2001-2003) Web: http://q12.org/ode/
19. K. Stoy, W.-M. Shen and P.M. Will, A simple approach to the control of locomotion in self-reconfigurable robots, Robotics and Autonomous Systems, Vol.44, No.3, pp.191-200, 2003
20. I. Tanev, DOM/XML-Based Portable Genetic Representation of Morphology, Behavior and Communication Abilities of Evolvable Agents, In Proceedings of the 8th International Symposium on Artificial Life and Robotics (AROB'03), Beppu, Japan, pp. 185-188, 2003

21. Y. Zhang, M. H. Yim, C. Eldershaw, D. G. Duff, K. D. Roufas, Phase automata: a programming model of locomotion gaits for scalable chain-type modular robots, IEEE/RSJ International Conference on Intelligent Robots and Systems (IROS 2003), October 27 - 31, Las Vegas, NV, 2003

3

Evolution of Khepera Robotic Controllers with Hierarchical Genetic Programming Techniques

Marcin L. Pilat[1] and Franz Oppacher[2]

[1] Department of Computer Science, University of Calgary,
2500 University Drive N.W.,
Calgary, AB, T2N 1N4, Canada,
pilat@cpsc.ucalgary.ca, http://www.pilat.org

[2] School of Computer Science, Carleton University,
1125 Colonel By Drive,
Ottawa, ON, K1S 5B6, Canada
oppacher@scs.carleton.ca, http://www.scs.carleton.ca/~oppacher

In this chapter, we evolve robotic controllers for a miniature mobile Khepera robot. We are concerned with control tasks for obstacle avoidance, wall following, and light avoidance. Robotic controllers are evolved through canonical GP implementation, linear genome GP system, and hierarchical GP methods (Automatically Defined Functions, Module Acquisition, Adaptive Representation through Learning). We compare the different evolutionary strategies based on their performance in evolution of robotic controllers. Experiments are performed on the Khepera GP Simulator for Windows. We develop the simulator as a user and developer friendly software to study GP and other robot controllers.

3.1 Introduction

Evolutionary computation studies how theories of evolution can be used to solve computational problems. Various evolutionary computation approaches currently exist with different methodologies and applications. We are interested in the area of genetic programming which uses evolutionary ideas to evolve computer programs.

robotics focuses on building machines to improve the lives of humans. Robots are designed to perform repetitive or dangerous tasks with excellent precision and dependability. However, robots require directions and programming to accomplish their goals.

In this chapter, we study the application of genetic programming techniques to the evolution of control programs for an autonomous miniature robot. We also present a software simulator for the Khepera miniature robot designed to study genetic programming based robotic controllers.

3.2 Genetic Programming

Genetic Programming (GP) was introduced by Koza [9] as an extension to genetic algorithms in order to enrich the chromosome representation. Instead of fixed-length strings, GP evolves pieces of code written over a specified alphabet consisting of a set of functions and a set of terminals. The chromosome encoding can be directly executed by the system or can be compiled or interpreted to produce machine executable code.

The main problem with genetic programming lies in its scalability. Genetic programming has been demonstrated to solve a variety of applications [11, 13] but it appears to lose its effectiveness for more complex real-world problems [5]. When we solve complex problems, we typically break the task into simpler sub-tasks and solve each sub-task. In contrast, regular GP tries to compute the entire solution to the problem at once. While this method is suitable for smaller problems, it is often not powerful enough to solve difficult problems. The problem decomposition technique of breaking down the task and solving its sub-tasks (called modularization) seems to be the right solution to overcome the complexity threshold of real-world problems.

Modularization techniques have been developed for GP but have generally employed a fixed decomposition structure provided by the experimenter. Hierarchical Genetic Programming (HGP) introduces modularization techniques to the GP system so that the GP can evolve module solutions to problems without human-imposed structure. This automatic modularization technique should improve the performance of genetic programming on difficult problems.

Koza [11] identifies five techniques that can enable hierarchical problem solving to reduce the effort needed to solve a problem: hierarchical decomposition, recursive application, identical reuse, parameterized reuse, and abstraction. Hierarchical decomposition is the act of breaking a problem into smaller sub-problems, solving the sub-problems, and combining their solutions into a solution for the problem. Recursive application of hierarchical decomposition to a problem is able to recursively break the problem down into small sub-problems that would be easy to solve by the system. Identical reuse is the process of using previously computed solutions to identical sub-problems, while parameterized reuse offers a way of applying the same problem solving mechanism to similar sub-problems via parameters. Abstraction deals with exclusion of irrelevant data from the problem environment.

Several hierarchical genetic programming methods have been suggested, each with its own advantages and disadvantages. The methods have been tested on various problems; however, current research does not adequately explain whether the studied HGP methods can, in general, outperform standard GP. In our research, we study the HGP methods: Automatically Defined Functions (ADFs) [11], Module Acquisition (MA) [2], Adaptive Representation (AR) [25].

3.3 Robotic Control

Programming robots by humans can be a difficult endeavor and is not well suitable for complex real-world applications. The area of evolutionary robotics deals with automatic generation of control programs for robots using evolutionary techniques.

The area of robotic control is often subdivided into three sub-areas: reactive, behavior-based, and hybrid [3]. Reactive control uses a simple set of condition-action pairs that define how the robot reacts to a stimulus. Brooks [6] proposed a multi-layer subsumption architecture where higher-level layers can subsume and block lower-level layers from action. Behavior-based architecture [14] uses a collection of interacting behaviours that can take input from the robot's environment sensors or other behaviours and produce output to the robot's effectors or other behaviours. Hybrid control strategies exist that offer a compromise between purely-reactive and behavior-based strategies.

Brooks [7] introduced the idea of using Artificial Life techniques to evolve control programs for mobile robots. Although no experimental results were presented, Brooks identified genetic programming as a hopeful technique for control program evolution. Koza [10] presented results of using GP to evolve emergent wall following behavior for an autonomous mobile robot. The control program was based on the subsumption architecture and demonstrated that GP can evolve control programs for mobile robots. In [12], Koza and Rice demonstrated that genetic programming can automatically create a control program to perform a box moving task. The paper also offered a good comparison between GP techniques and reinforcement learning techniques in accomplishing the task.

Reynolds [22] has used genetic programming to evolve a controller program for tiny critters in a simulated environment. The critter tasks were to manoeuver in a static obstacle environment (obstacle-avoidance) and avoid a predator. In this ALife-inspired predator-pray paradigm, the fitness criteria was based on the sum of the critter lifetimes. Results showed interesting partial solutions to the task but failed to show herding behavior such as observed in animals.

Nordin and Banzhaf [16, 19, 17, 20, 18] have experimented with a simulated and real Khepera miniature robot to evolve control programs using genetic programming. They used the Compiling Genetic Programming System (CGPS) [15] which worked with a variable length linear genome composed of machine code instructions. The system evolved machine code that was directly run on the robot without the need of an interpreter.

The initial experiments of Nordin and Banzhaf [16, 17] were based on a memory-less genetic programming system with reactive control of the robot. The system performed a type of symbolic regression to evolve a control program that would provide the robot with 2 motor values from an input of 8 (or more) sensor values. GP successfully evolved control programs for simple control tasks such as: obstacle avoidance, wall following, and light-seeking. The work was extended [20, 21] to include memory of previous actions and a two-fold system architecture composed of a planning process and a learning process. Speed improvements over the memory-less system were observed in the memory-based system and the robots exhibited more complex behaviours [20]. Summary of the techniques used and tasks studied can be found in [4].

We are interested in the reactive control of a Khepera robot using genetic programming techniques. In reactive control experiments, robots learn while travelling through the experimental environment. No separate fitness cases are used to calculate fitness and thus the robot positions do not need to be reset for the purpose of fitness calculation. The reactive control problem is difficult since it requires dynamic fitness function evaluation where the individual fitness values depend on the local environment of the robot. However, the problem presents a more realistic dynamic learning environment.

The learning method used in an evolutionary algorithm can greatly influence the successfulness of the solution to the problem. Due to their beneficial properties, we feel that hierarchical genetic programming methods will offer advantages to the problem of reactive robotic control.

3.4 Khepera Simulators

Robotic simulators play an important role in robotic experimentation. Robotic equipment can be costly and requires proper facilities. Software simulators offer the experimenter a test-bed for robotic technologies when a physical robot cannot be acquired. Some simulators provide a very accurate model of the environment and of interactions in the environment. Such simulators can be used as valid substitutions for real robots for testing various robotic tasks.

Some robotic research on physical robots requires constant supervision and periodical rearrangement of the robots within the environment. For such research, robotic simulators have an advantage to physical robots. Simulators can be left unsupervised and can be programmed to automatically perform human actions such as relocation of robots in the simulated environment. This can considerably speed up experimentation time and requires less human time.

The main disadvantage of software simulators is the inexact model of the environment. A real physical environment contains noisy data that can greatly influence the results of an experiment. One of the goals of using a robotic software simulator is to be able to reproduce similar results on the physical robot. Thus, the software environment must contain noise comparable to the real physical environment.

3.4.1 Khepera Robot

The Khepera robot is a miniature mobile robot created and sold by K-Team S.A. (http://www.k-team.com) - a Swiss company specializing in development and manufacture of mobile mini-robots. Recently, K-Team has created a new Khepera II robot with an improved micro-processor, more memory, and a wider range of capabilities. Our research is based on the original Khepera robot.

Khepera is circular, with a diameter of 55mm and height of 30mm. The robot can sense its environment with 8 built-in infra-red proximity and ambient light sensors. Two motors with controllable acceleration are used to move the robot in the environment. Fig. 3.1 provides a schematic diagram of the robot's sensors and motors.

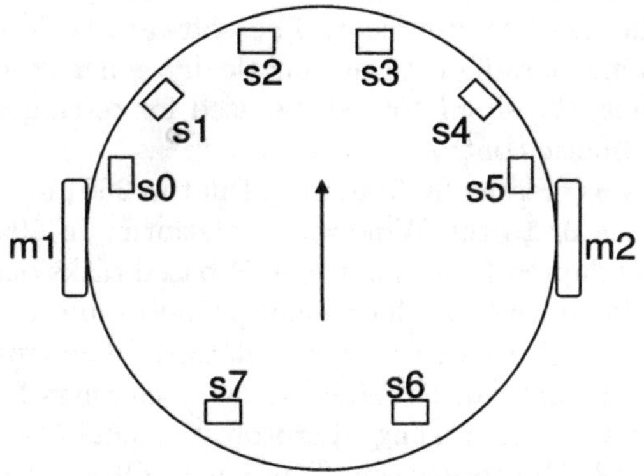

Fig. 3.1. Schematic view of the Khepera robot. Sensors are labelled s0 to s7 and motors are labelled m1 and m2.

The brain of the Khepera robot is a 16Mhz Motorola 68331 micro-processor with 256 KB of RAM and 128-256 KB of reprogrammable ROM memory. The ROM contains a simple operating system and communication interface to a host computer. The robot can execute its own programming that can be either provided through a serial connection or downloaded into the onboard memory.

The Khepera robot can be equipped with a variety of extension turrets that provide it with abilities to perform more complex tasks. Some extension turrets are: gripper turret used for object recognition and manipulation, video turret for on-board camera ability, and I/O turrets for improved communication with the host computer.

3.4.2 Khepera Simulator

The original Khepera Simulator (`http://diwww.epfl.ch/lami/team/michel/khep-sim`) was developed by Olivier Michel at the Microprocessor Systems Lab (LAMI) of the Swiss Federal Institute of Technology (EPFL). The latest version of the simulator (version 2.0) is available free-of-charge for research use and it is written exclusively for the UNIX® platform.

Many other software simulators for the Khepera robots are currently available. Cyberbotics (http://www.cyberbotics.com) specializes in development of 3D simulation software for mobile robots. The software - Webots - supports a variety of robots rendered in a 3-dimensional environment.

3.4.3 Khepera GP Simulator

The Khepera GP Simulator for Windows® is a software package to simulate Khepera robots in their environment. The software is designed to use the genetic programming paradigm to automatically generate control programs for the robots. Thus, the simulator can be used for testing of GP techniques in the domain of robotic control.

The simulator was created by Marcin L. Pilat in 2001 as a port of the original Khepera Simulator to the Windows® platform. In 2003, the simulator was improved and adapted for simulating GP-based tasks on Khepera robots. Version 3.0 is available free for educational purposes and can be downloaded from the author's website (http:// www.pilat.org/ khepgpsim). The source code is also available and can be modified by researchers for specific experiments. The code was written using Microsoft® Visual C++®, Microsoft® Foundation Class (MFC) Library, and Component Object Model (COM). The simulator is only available for the Windows® platform.

The main purpose of the Khepera GP Simulator is to simulate a physical Khepera robot in its environment. The simulation includes sensing of the environment using the robotic sensors and interacting with the environment using the robotic actuators (motors powering the wheels). Noise is added to the simulation in order to approximate a noisy environment. Multiple Khepera robots can be simulated together thus allowing the study of more complex robotic behaviours requiring interaction between the robots (e.g. box-pushing, following, collective garbage collection).

The environment of the robot is modeled as a scalable rectangular working area. All items in the environment are treated as objects. There are three types of objects - building objects (bricks, corks, boxes), light objects (lamps, light boxes), and robot objects. Robot objects are simulated Khepera robots placed in the environment. Any object can be manipulated in the environment in real-time during a simulation run.

The Khepera GP Simulator is specifically designed to study GP-based robotic controllers but can be easily adapted to non-GP controllers. The controller dictates the actions of the robot in the environment. The GP controllers included with the simulator modify a population of robotic control programs in order to evolve certain tasks (or behaviours).

The robotic controller provides a set of motor values to be used by a robot during each step of the simulation. The motor values are processed to yield a force vector that specifies the direction of the motion and the amount of force the robot applies in the world. The force vector is then used to calculate the next position and rotation of the robot. Collisions are handled by a simple vector-based collision engine with modifiable parameters.

Each learning task (such as obstacle avoidance, wall following) can be evolved with any type of GP controller. The controller type specifies the chromosome structure and chromosome interactions during evolution. Multiple tasks can be evolved by the same GP controller type with different specifications of the chromosome structure. A task contains a population of chromosomes; thus, it can be used to store snapshots of the population during evolution.

Each task contains a fitness function which provides guidelines for the evolution of the population of control programs. The fitness function can be thought as a formal definition of the learning task. Fitness functions in the simulator are dynamic and can be easily modified at runtime. The fitness function definitions are written using a scripting language - Microsoft® JScript™. This scripting language is based on Java™and is available free-of-charge from Microsoft® Corporation. JScript™provides the user with a rich scripting language to define the fitness function. The language supports a variety of pre-defined functions and the ability to create variables.

The GP controllers in the simulator gather statistical information during the run of the evolutionary algorithm. This information is stored in order to analyze the performance of an evolutionary run. For each generation, the statistics engine stores average and best population fitness, robotic collisions, chromosome complexity, and population entropy values. Population entropy [23] measures the state of a dynamic system represented by the population and can be correlated with the state of population diversity.

Complexity of the chromosomes in the population is stored using three complexity measures: size, structural complexity, and evolutional complexity [27]. The size measure specifies the raw size of the chromosomes defined as the number of instructions in a linear genome chromosome or the number of tree nodes in the tree-based chromosome representation. The structural complexity measure includes the sizes of all unique function trees called from within an individual. Evaluational complexity of an individual is measured recursively and includes sizes of all function trees embedded in the individual. This measure approximates the number of computational units required for execution of the individual program.

3.5 Robotic Controllers

Our research into robotic controllers builds on research done by Nordin and Banzhaf [20] to evolve GP robotic controllers for the Khepera robot. Nordin and Banzhaf were able to evolve controllers for various learning tasks (such as obstacle avoidance and wall following). In our research, we compare the linear genome GP method they have used in their experiments to canonical tree-based GP representation and three most popular Hierarchical Genetic Programming methods: Automatically Defined Functions, Module Acquisition, and Adaptive Representation.

The GP system in the robotic controller evolves control programs that best approximate a desired solution to a pre-defined problem. This procedure of inducing a symbolic function to fit a specified set of data is called Symbolic Regression [18]. The goal of the system is to approximate the function:

$$f(s0, s1, s2, s3, s4, s5, s6, s7) = \{m1, m2\} \tag{3.1}$$

where the function input is the robotic sensor data (*s0-s7*) and the output is the speed of the motors controlling the motion of the robot (*m1-m2*). The control program code of each individual constitutes the body of the function. The results are compared using a behaviour-based fitness function that measures the accuracy of the approximation by the deviation from desired behavior of the robot.

In our research, we deal with a population of control programs for the Khepera robot. A steady-state tournament selection GP algorithm is applied to the population in order to evolve control programs that accomplish the specified learning tasks.

3.5.1 Tree-based GP

The canonical GP implementation uses a tree-based chromosome representation [9, 11]. The chromosome (originally coded as a LISP S-expression) represents a parse-tree that can be easily transformed into machine code. The internal nodes of the program tree are chosen from a set of parameterized functions with parameters as subtrees. Leaf nodes are chosen from the set of parameter-less functions and terminals. The terminal set is usually composed of variables and constants. Variables are place holders in the chromosome that are filled in with values during execution. Functions perform calculations or actions and can optionally have parameters. To generate tree-based chromosomes, we use the function and terminal sets as shown in Table 3.1.

Program trees of each individual are created in a recursive manner. Three methods have been suggested for the creation of the initial random population: *full*, *grow*, or *ramped half-and-half* [9]. The *full* method creates trees with all leaf nodes at equal depth and is the method used in our implementation. The

Table 3.1. Contents of the function set and terminal set used by tree-based chromosome representations.

Function Set:	`Add, Sub, Mul, Div, AND, OR, XOR, <<, >>, IFLTE`
Terminal Set:	`[s0-s7]` (8 proximity sensors),
	`[l0-l7]` (8 ambient light sensors),
	`[0-8192]` (constants in given range)

grow method grows trees of variable size and the *ramped half-and-half* method creates a mixture of trees with different heights through either the *full* or the *grow* method.

Two genetic operators are used in the tree-based chromosome representation: reproduction and crossover. Reproduction copies a chromosome into the next generation. Single subtree switching crossover is applied to the two fittest individuals in a tournament, with a given probability. We use a crossover probability of 0.9 in all tree-based chromosome representation experiments.

3.5.2 Linear Genome GP

Nordin [15] provided a linear genome GP system which stores 32-bit instructions that can be executed directly on a processor. Nordin claimed the execution speed of the Compiling Genetic Programming System (CGPS) is several orders of magnitude faster than of an equivalent interpreted tree-based GP system [15]. The major disadvantage of the CGPS system is that it is only usable on a processor supporting the specific machine-code instruction set used. To be used on a processor with a different instruction set, the system needs to be either rewritten or interpreted. The CGPS was later called the Automatic Induction of Machine code by Genetic Programming (AIMGP) system [21].

The linear genome method was applied by Nordin and Banzhaf [18] to evolve a robotic controller for Khepera robots. The structure of our linear genome GP controller closely resembles the controller used by Nordin and Banzhaf. We represent each instruction as a text string and process it through a genome interpreter prior to evaluation. This encoding improves the readability of the program code compared to the binary approach of Nordin and Banzhaf but suffers a loss in performance due to processing of the string based instructions. However, the performance of the string-based representation is sufficient for the purpose of our research.

In the linear genome GP system, each individual is composed of a series of instructions (genes). The instructions are of the following format:

$$resvar = var1 \;\; op \;\; \{var2|const\} \tag{3.2}$$

where *resvar* is the result variable and *op* is a binary operator working on either two variables (*var1* and *var2*) or a variable and a constant (*var1* and *const*).

Each individual is randomly assigned a height (number of instructions) from 1 to the maximum specified height. For each part of an instruction, a value is selected randomly from a set of primitive values. Table 3.2 provides primitive value sets of the instruction parts used in our linear genome experiments.

Table 3.2. Primitive values of instruction parts in the linear genome GP method.

Part	Primitive Value Set
res	motor values ($m1, m2$) intermediate variables ($a - f$)
var1, var2	proximity sensor values ($s0 - s7$) light sensor values ($l0 - l7$) intermediate values ($a - f$)
op	add (`+`), subtract (`-`), multiply (`*`), left shift (`SHL`) right shift (`SHR`), XOR (`^`), OR (`|`), AND (`&`)
const	integer value in range: 0-8191

The linear genome GP method employs three genetic operators: reproduction, crossover and mutation. The crossover operator uses a simple variable length 2-point crossover applied to the list of instructions (genes) of two fittest individuals of a tournament. Genes are treated as atomic units by the crossover operator and are not modified internally. Simulated bit-wise mutation modifies the contents of a gene. Crossover probability of 0.9 and mutation probability of 0.05 are used in the linear genome experiments.

3.5.3 Automatically Defined Functions HGP

Koza's Automatically Defined Functions (ADFs) [11] method is the oldest and most widely used HGP method. The method automatically evolves function definitions while evolving the main GP program that is capable of calling the functions. The ADF HGP method implemented in our research is based on the ADF method proposed by Koza.

The ADF method has been demonstrated to be advantageous in solving more complex versions of problems than possible by standard GP (e.g. 6-parity problem) [11]. The major disadvantage to the method is that the user must specify the structure of the ADF chromosomes (number of functions and arguments) and the function and terminal sets required by each function. In a true automatic HGP system, this type of information should be evolved by the GP rather than provided by the user. Taking the downside of ADFs into consideration, current research is centered around operations that automatically modify the structure of the ADF chromosome and the number of ADFs [13].

The method is an extension of the tree-based GP method and shares its basic structure. An ADF chromosome consists of two distinct parts: the function defining branch, and the result producing branch. The function defining part is composed of one or more ADF definition branches which describe the structure of each ADF. The result producing branch contains the code of the resulting program. This code can call any function defined in the function defining branch of the same chromosome. Invariant nodes are fixed structural nodes and are present in every ADF chromosome. Non-invariant nodes define the bodies of the ADF definitions and the result producing branch and are modified during evolution.

All ADFs defined in an individual are available locally to the program tree of the same individual. The number of ADFs present in each chromosome and the number of arguments for each ADF are specified as parameters. We use chromosomes with one, two, and three ADF definitions and two function arguments. Zero or multiple ADFs can be called from within the result producing branch. Some recursive ADF implementations allow calling of ADFs from within other ADFs. This leads to problems with circular evocation of ADFs and requires extra protection. Due to the increase of implementation complexity, we do not allow ADF calls inside ADF definitions in our implementation.

The result producing branch is built using a standard terminal set and standard function set (shown in Table 3.1) augmented with the ADFs contained in the same chromosome. Separate terminal and function sets are used by the function defining branches to define the ADFs. The ADF branch function set is identical to that of the tree-based chromosome representation whereas the ADF terminal set is composed of ADF argument variables and constants.

Tree-based reproduction and crossover genetic operators are used in the ADF chromosomes. The crossover operator can only swap non-invariant nodes of the same type using branch typing [11].

3.5.4 Module Acquisition HGP

The Module Acquisition (MA) method of Angeline and Pollack [2] employs two new operators of compression and expansion to modularize the program code into subroutines. The subroutines contained in the subroutine collection are frozen in time and cannot be modified during evolution of the program trees. The Module Acquisition method automatically generates a hierarchical module structure [1]; however, no clear advantages of the method have yet been provided. Kinnear has compared MA to ADFs on the even-4-parity problem [8] and concluded that the method does not offer improvement in space or time over the ADF method.

The chromosome structure is identical to that of the original tree-based chromosomes with standard tree-based function and terminal sets. Modules (subroutines) are created locally for each chromosome from subtrees of the program tree and propagate through the population solely by reproduction and crossover. Module nesting is allowed inside program trees of other modules; however, by the nature of their creation, modules are not recursive.

The Module Acquisition method employs four genetic operators: reproduction, crossover, and two mutation operators of compression and expansion. The reproduction and crossover operators perform as for tree-based chromosomes. The compression operator creates a new subroutine from a randomly selected subtree of an individual in the population using depth compression [1]. We use a maximum depth value from range 2-5. Branches beyond the maximum depth are used as parameters to the new subroutine.

Since the compression operator lowers the diversity of the population by removing subtrees, an expansion operator is also provided to counteract the negative effects. The expansion operator reverses the process of the compression operator by substituting the original subtree for a subroutine call in the chromosome tree. The subroutine is removed from the module list of the chromosome if it is no longer used.

The special mutation operators are applied after the standard tree-based reproduction and crossover operators. We set the probability of compression to 0.1 and probability of expansion to 0.01.

3.5.5 Adaptive Representation HGP

Rosca and Ballard proposed the Adaptive Representation method to dynamically extend the function set with identified building blocks [25]. The method uses standard tree-based representation and searches for blocks of code (defined as subtrees of a given maximum height). Blocks are parameterized into functions by substituting each occurrence of a terminal by a variable. Unlike in the ADF HGP approach, the functions are discovered automatically and without human-imposed structure. The method differs from the MA HGP approach by the algorithms used in function discovery and management of the function library. Our implementation of the AR method is based on the improved Adaptive Representation through Learning (ARL) algorithm [26].

The method works by incrementally checking the population for fit building blocks. Block fitness is dependant on the performance of the individual where the block resides (and, thus, the block) or the performance of a part of the individual (e.g. using a block fitness function). Evolution is done in epochs which are defined as sequences of consecutive generations where no fit building blocks are discovered. At the end of each epoch (i.e. after a discovery of a candidate building block) a proportion of the population (constituting the lowest performing individuals) is replaced by individuals that are randomly generated from the new extended function set. Rosca and Ballard provide theoretical discussion on the usefulness of their approach in improving the speed of evolution over standard GP [25]. It is unclear, however, how to discover candidate building blocks without additional domain knowledge.

The structure of the ARL chromosome program trees is identical to that of the tree-based GP method. The function set is dynamically extended by the evolutionary algorithm through creation of new functions. Nesting of functions is allowed; however, recursive function calls are not possible due to the function creation method.

The main advantage of the ARL algorithm is the automatic discovery of useful subroutines through the concepts of differential fitness and block activation [24]. Differential fitness is defined as the difference in fitness between an individual and its least fit parent. Rosca states that large differential fitness can be the result of useful combinations of blocks of code in the individual [24]. Block activation is defined as the number of times a block of code is executed during evaluations of the individual. Rosca states that only blocks with high block activation values should be considered candidate blocks. We do not implement the concept of block activation because of the large performance overhead on the system.

In our implementation of the ARL algorithm, we select the most promising individual (based on differential fitness) from the set of promising individuals discovered during the last generation. Candidate blocks of small height (tree height of 3) are chosen from the most promising individual. The blocks are generalized into subroutines which extend the function set.

Rosca [24] computes subroutine utility which is analogous to schema fitness for subroutines. The utility is defined as the accumulation of rewards for a subroutine over a fixed time window and is calculated by a special utility function. Using subroutine utility, low performing subroutines are removed from the function set. We implement a simpler measure of subroutine utility by assigning to each subroutine an integer utility value denoting the number of generations until an unused subroutine is removed from the function set. Utility value of each unused subroutine is decremented each generation until it reaches 0 and the subroutine is removed from the population.

The run of the ARL algorithm is divided into epochs which were defined as sequences of consecutive generations in which no new candidate building blocks are discovered [25]. The ARL algorithm provides a concrete definition of epoch creation using population entropy [23] which provides a measure of the state of a dynamic system represented by the population. Rosca [23] compares the population-based dynamic system to a physical or informational system with similar behavior.

In our implementation, entropy is measured by grouping individuals of the population into a set of classes based on their behavior (phenotype). Shannon's formula is then used to calculate the entropy:

$$E(P) = -\sum_{k} p_k \cdot \log p_k \tag{3.3}$$

where p_k is the proportion of the population P grouped into partition k. Entropy is usually computed based on raw individual fitness; however, we could not use raw fitness because of the dynamic nature of our fitness calculation.

We compute a standardized fitness measure through an average of three fixed test cases. Each test case provides resulting robotic motor values from individual evaluation on a fixed set of input sensors. We then partition the individuals into 20 categories based on their standardized fitness measures. The population entropy value is calculated by applying Shannon's formula on the partition categories.

The measure of population entropy is important since it correlates to the state of diversity in the population during a GP run. Drops in population entropy signify drops in population diversity. The ARL method tries to counteract the drops in population entropy by creation of new individuals. The start of a new epoch is decided using a static entropy threshold of 1.5. New epoch begins and subroutines are discovered when the entropy value of the population falls below the threshold.

After the discovery of new subroutines, the function set is extended by the new functions. The ARL method generates random individuals using the new function set. The new individuals replace a fixed proportion of the worst performing individuals in the population. We use a replacement fraction of 0.2 in our experiments. Genetic operators of reproduction and crossover are similar as for the tree-based method.

3.6 Results

3.6.1 Obstacle Avoidance

The task of obstacle avoidance is important for many real-world robotic applications. Robotic exploratory behavior requires some degree of obstacle avoidance to detect and manoeuver around obstacles in the environment. We define obstacle avoidance as robotic behavior steering the robot away from obstacles in the testing environment. For the Khepera robot, this task is equivalent to minimizing the values of the proximity sensors while moving in the environment.

We select a fitness function based on the work of Banzhaf et al. [4]. The function is composed of two opposite parts: pain and pleasure. The pleasure part is computed from motor values and encourages the robot to move in the environment using straight motion. The pain part is composed of sensor values and punishes the robot for object proximity. The fitness function can be expressed as an equation:

$$Fitness = \alpha(|m_1| + |m_2| - |m_1 - m_2|) - \beta \sum_{i=0}^{7} s_i \tag{3.4}$$

where m_1 and m_2 are motor values and s_0 to s_7 are proximity sensor values. The value of α is set to 10 and value of β to 1. Parameter values were chosen based on tuning experiments.

Various robotic behaviours are observed while learning the obstacle avoidance task. We subdivide the learned behaviours into groups based on the complexity and success rate of each behavior. The simplest (Type 1) behaviours are solely based on the blind movement of the robot (straight, backup, curved). The second level (Type 2) of behaviour (circling, bouncing, forward-backup) includes behaviours with noticeable use of sensor data. The highest level (Type 3) of behaviour is called sniffing and demonstrates obstacle detection and avoidance. The perfect sniffing behaviour involves obstacle sniffing and straight motion behaviours that combine into smooth obstacle avoidance motion around the entire testing environment. Summary of the observed behaviours is provided in Fig. 3.2.

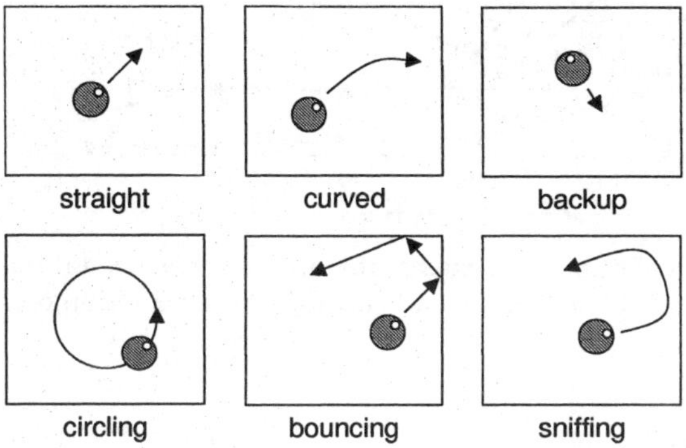

Fig. 3.2. Summary of behaviours learned during experimentation with the Khepera robot.

In our analysis of method performance, we examine population entropy stability, average chromosome complexity stability, and average generation of initial behaviour occurrence. Entropy and complexity stability is defined as a gradual change of the measured values over time without large abrupt value changes.

For the obstacle avoidance task, the representation method with the most stable entropy values is the ARL method. The linear genome and ADF methods also provide long, stable entropy values but with larger variations. The MA and tree-based representations provide the worst stability with large drops of entropy values. Most stability in the average chromosome size values is seen with the linear genome method. Among the HGP methods, the most stable complexity measures are seen with the ARL method and least stable with the MA method.

Type 2 and 3 behaviours are analyzed to calculate average generation values of first occurrence of the behaviours. We do not take into consideration Type 1 behaviours since they are not directly applicable to the studied task. Summary of the results of our behaviour calculation can be found in Fig.

3.3. The method with best (smallest) values is the ARL HGP method and with worst (largest) values is the linear genome GP method. Overall, the HGP methods perform comparable to the tree-based GP method. Trace run of perfect evolved obstacle avoidance behaviour is shown in Fig. 3.4.

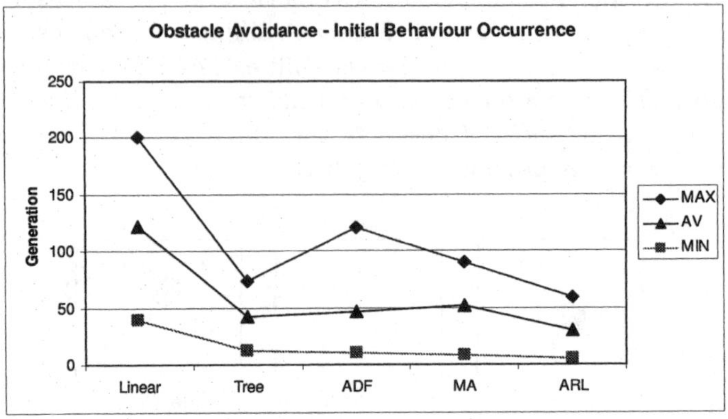

Fig. 3.3. Graphs of minimum, maximum, and average generations of first detection of Type 2 and 3 obstacle avoidance behaviour for each chromosome representation method.

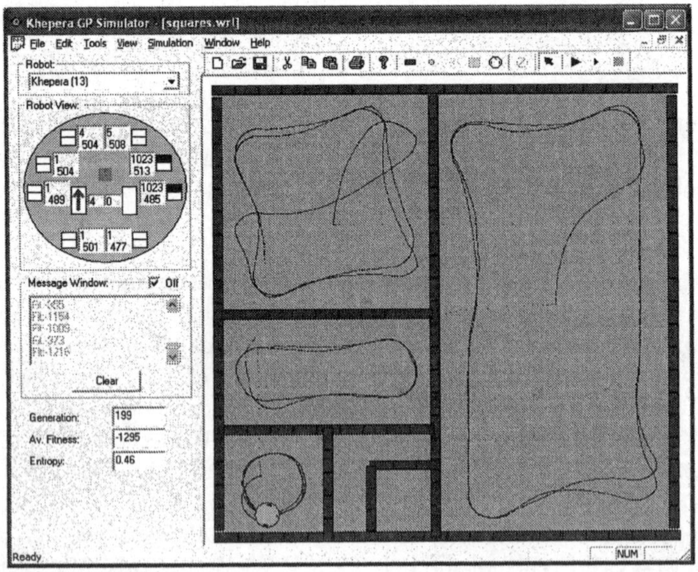

Fig. 3.4. Trace runs of perfect evolved obstacle avoidance behaviour in various testing environments.

3.6.2 Wall Following

The task of wall following allows the robot to perform more difficult and interesting behaviours such as maze navigation. The purpose of the wall following task is to teach the robot to walk around the boundaries of obstacles with a certain desirable distance away from the obstacles. The learned task should include some obstacle avoidance behaviour; however, that is not the main requirement of the experiments.

The wall following fitness function is composed of a sensor part and a motor part. The sensor part computes a sensor value from a subset of the robotic sensor values. The motor part is calculated by computing an absolute motor sum minus the absolute value of the difference. The fitness function is provided in Fig. 3.1. In our experiments, we set the values for the free parameters of the fitness function as follows: $\forall_i a_i = 1, \alpha = 100, \beta = 1$. Parameter values and fitness function definition were chosen based on tuning experiments.

Algorithm 3.1 Wall Following Fitness Function

input: Left $= a_0 \cdot s_0 + a_1 \cdot s_1 + a_2 \cdot s_2$,
Right $= a_5 \cdot s_5 + a_4 \cdot s_4 + a_3 \cdot s_3$,
MotorPart $= |m_1| + |m_2| - |m_1 - m_2|$;
output: Fitness;

```
 if (Right ¿ 1023)
   RightSensorPart = 1000 - Right;
 else if (Right ¡ 20)
   RightSensorPart = (1000/20) * Right;
else
   RightSensorPart = 1000;
if (Left ¿ 1023)
   LeftSensorPart = 1000 - Left;
 else if (Left ¡ 20)
   LeftSensorPart = (1000/20) * left;
 else
   LeftSensorPart = 1000;
 Fitness = α· MotorPart + β· (RightSensorPart + LeftSensorPart);
```

Only six sensors ($s_0 - s_5$) are used in calculating the sensor part of the fitness calculation. The sensors represent the side and front sensors of the robot. The calculated sensor part value acts as either pleasure or pain depending on the sensor values. The robot is punished when it is either too far away from a wall or too close to it. The training environment consists of a long, straight stretch of corridor and curved environment boundaries.

Summary of observed behaviours is provided in Fig. 3.2. We partition the behaviours into categories based on their relative performance and success. The Type 1 category is of poor wall following behaviour and consists of simple wall-bouncing and circling behaviours. The Type 2 category of good wall following behaviour contains wall-sniffing and some maze following. The best behaviour category, Type 3, consists of perfect maze following behaviour without wall touching.

The most stable entropy is noticed in experiments using the ARL HGP method. The least stable entropy is observed using the ADF method and includes a large initial drop of entropy values to a low, stable level. Good stability of average size values is seen in the linear genome GP and ARL HGP methods. The largest drops in average chromosome size are noticed with the MA method.

We calculate average generation values of first occurrence of Type 2 and 3 behaviours. Type 1 category bahaviour is not directly applicable to the studied task. Summary of our behaviour calculation can be found in Fig. 3.5. The ARL method produces the best average results with smallest deviation whereas the worst performance is seen using the MA method. Trace run of perfect evolved maze-following behaviour is shown in Fig. 3.6.

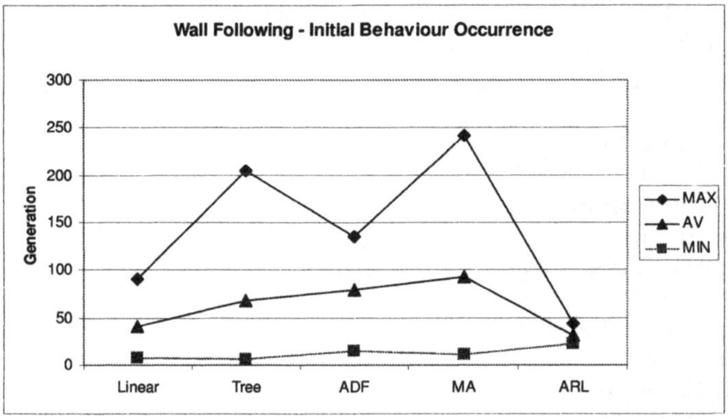

Fig. 3.5. Graphs of minimum, maximum, and average generations of first detection of Type 2 and 3 wall following behaviour for each chromosome representation method.

3.6.3 Light Avoidance

The light avoidance task is similar to the obstacle avoidance task but relies on the ambient light sensors of the robot instead of the proximity sensors. The source of light in the training and testing environments is composed of overhead lamps that cannot be touched by the robot. The robot must learn to stay inside an unlit section of the world environment while moving as much as possible.

The fitness function for light avoidance is derived from the fitness function for the obstacle avoidance task. The function contains a pleasure part

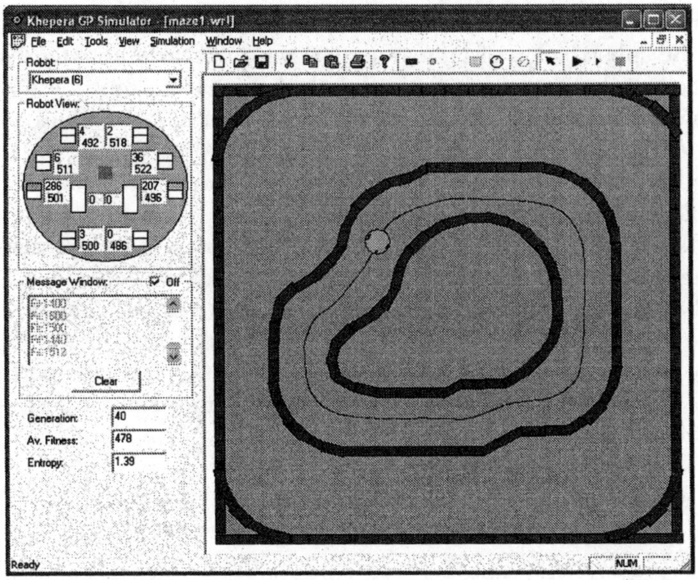

Fig. 3.6. Trace run of perfect evolved maze-following behaviour.

computed from the motor values of the robot and a pain part computed from the light sensors. Proximity sensors are not part of the fitness evaluation. A formal definition of the function is given as:

$$Fitness = \alpha(|m_1| + |m_2| - |m_1 - m_2|) - \beta(4000 - \sum_{i=0}^{7} l_i) \qquad (3.5)$$

where m_1 and m_2 are motor values and l_0 to l_7 are ambient light sensor values. We set a default value of 10 for α and default value of 1 for β in our experiments. Because of the definition of light sensor values (with 0 as maximum light and 500 as minimum), we subtract the sensor sum from 4000 (8 sensors of 500 value each) to make the fitness function behave similar to the fitness function for obstacle avoidance. Parameter values were chosen based on tuning experiments.

The training environment is composed of a rectangle of darkness surrounded by lights and a circular light island in the middle of the darkness area. The testing environment contains a similar dark rectangular area without the middle island.

We subdivide the learned behaviours of the robots into two categories. The Type 2 category of behaviour consists of circular, oval or uneven robot maneuvers with low degree of light detection and avoidance. Type 3 behaviour classifies definite light detection and avoidance behaviours. Perfect behaviour usually consists of travelling around the boundary of the dark area in the testing environment. Summary of possible behaviours can be found in Fig. 3.2 with the substitution of light boundaries for obstacle boundaries.

The most stable entropy is noticed with the linear genome method and the least stable with the ADF method. Most stable average size values are noticed using the ARL method. The linear genome and tree-based representations also provide quite stable average size behaviour. The worst average size stability is seen with the MA representation method.

Results with Type 2 and 3 light avoidance behaviour are processed to calculate average generation values of first occurrence of the behaviour. Summary of our behaviour calculation results can be found in Fig. 3.7. The best (lowest) values are from experiments using the ARL method while the worst (highest) values are from linear genome experiments. The HGP methods perform comparable to or better than the tree-based method. Trace runs of perfect evolved light avoidance behaviour are shown in Fig. 3.8.

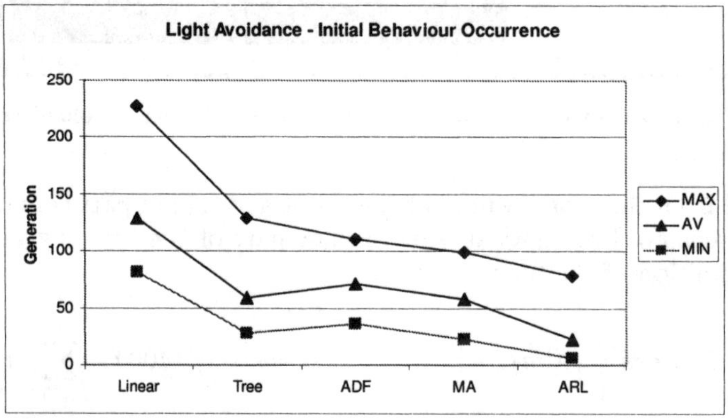

Fig. 3.7. Graphs of minimum, maximum, and average generations of first detection of Type 2 and 3 light avoidance behaviour for each chromosome representation method.

3.7 Summary

Our research deals with evolution of robotic controllers for the Khepera robot. We are interested in the population of individuals making up the robotic controller. The reactive robotic control problem provides a challenge to the genetic programming paradigm. With the lack of test cases for fitness function evaluation, the fitness of an individual can differ greatly depending on the immediate neighbourhood of the robot. The definition of the fitness function can influence the population contents and thus the resulting behaviours.

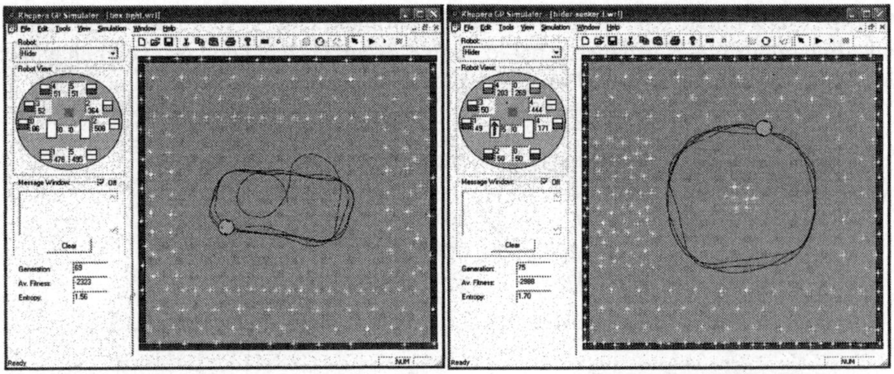

Fig. 3.8. Trace runs of perfect evolved light avoidance behaviour in different testing environments.

Robotic controllers often over-adapt to the training environment. This problem of overfitting is a common problem in genetic programming. A choice of proper training environment for a particular task is thus very important. From our obstacle avoidance and wall following task learning experiments, we notice that sharp corners of the environment form an area of difficulty for the robotic controller. This is probably caused by a corner fitting between the fields of view of the proximity sensors.

The population entropy value is an important indicator of population diversity in our experiments. Good trained behaviour is found in populations with relatively high entropy value (above 0.6). Low entropy value signifies convergence in the population which usually accompanies a convergence to a low average chromosome size. Populations of individuals with low chromosome size do not contain enough information to successfully search for a good solution. No special measures are taken to prevent bloating in our experiments; however, a maximum tree height (or maximum number of instructions for the linear genome method) is specified for each chromosome.

We examine three HGP learning methods: Automatically Defined Functions (ADF), Module Acquisition (MA), and Adaptive Representation through Learning (ARL) and two GP methods: tree-based and linear genome. Robotic controllers using each method are able to evolve some degree of proper behaviour for each learning task. Summary of method performance is available in Table 3.3. We treat the tree-based method as a basis for evaluating the performance of the linear genome and HGP methods. We define the behaviour of the tree-based method as average. Sample plots of population entropy and chromosome complexity observed in tree-based experiments are provided in Fig. 3.9.

The best entropy and best average size stability is seen with experiments using the ARL method (see Fig. 3.10). The worst entropy behaviour is seen mainly with the ADF method (as shown in Fig. 3.11) but also with the MA

Table 3.3. Summary of results from our experiments for each of the studied methods. Behavioural performance is based on first occurrence of good evolved behaviour. Methods are compared based on relative performance.

Method	Entropy Stability	Size Stability	Behavioural Performance
Linear GP	excellent	excellent	poor
Tree GP	average	average	average
ADF HGP	poor	average	average
MA HGP	average	poor	average
ARL HGP	excellent	excellent	excellent

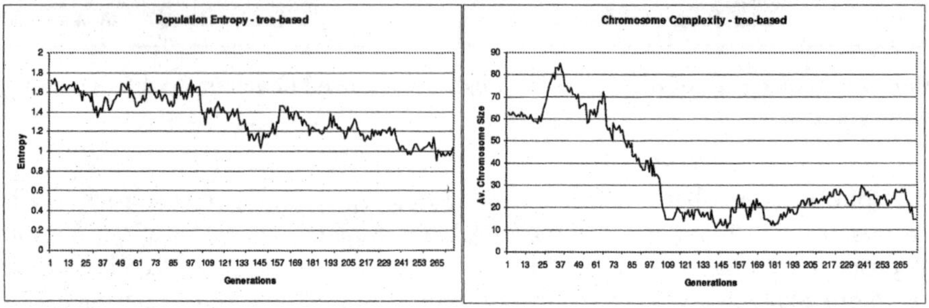

Fig. 3.9. Graphs of entropy and average size vs. the number of generations in a sample run using the tree-based method.

and tree-based methods. The worst average size behaviour is noticed with the MA method for all the studied tasks (see Fig. 3.12).

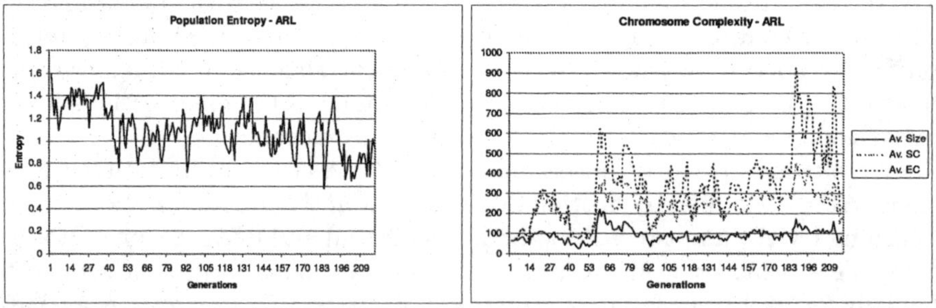

Fig. 3.10. Graphs of entropy, average size, average SC and average EC vs. the number of generations using the ARL method.

Throughout most of our experiments, the linear genome method enforces a stable level of entropy and average chromosome size (as seen in sample plot of Fig. 3.13). This behaviour is probably due to the different crossover operator in the linear genome method than in the tree-based methods and by the additional mutation operator. Because of the stable entropy levels,

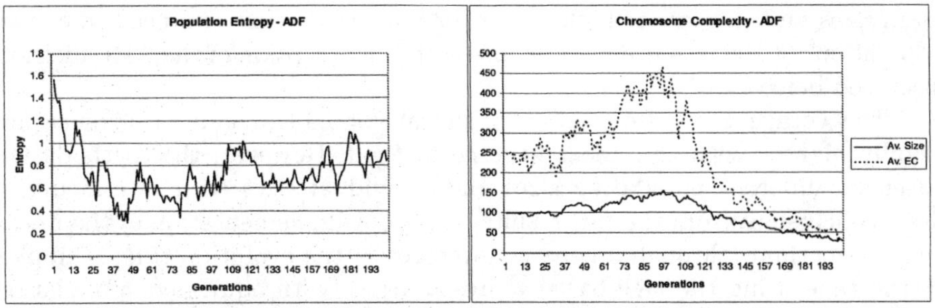

Fig. 3.11. Graphs of entropy, average size and average EC vs. the number of generations in a sample run using the ADF method.

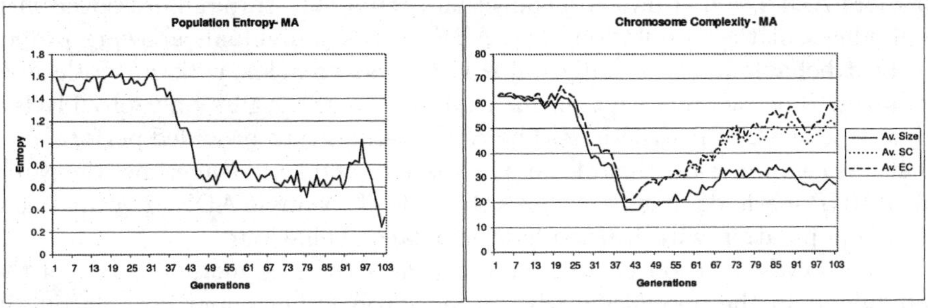

Fig. 3.12. Graphs of entropy, average size, average SC and average EC vs. the number of generations in a sample run using the MA method.

populations of 50 individuals are enough to provide stable behaviour for many generations.

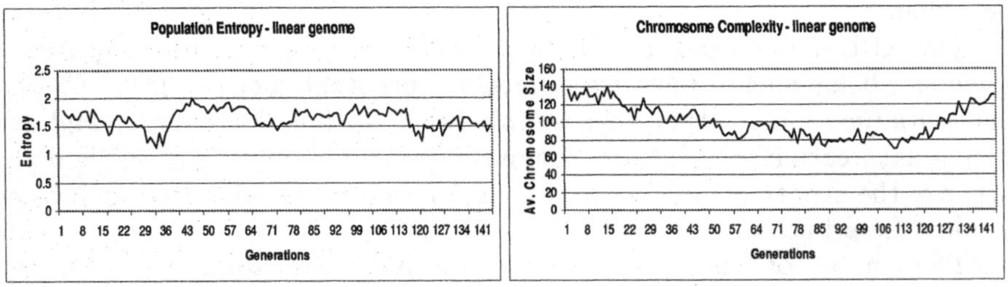

Fig. 3.13. Graphs of entropy, average size, average SC and average EC vs. the number of generations using the linear genome method.

With the tree-based chromosome representations, entropy value stability depends on the definition and parameter values of the fitness function. Tendency toward smaller program sizes is seen with the half-and-half chromosome creation method or the full method with small sized populations. To keep program sizes and entropy at high values for reasonable time, we need to evolve populations of 100 or more individuals (with the exception of the ARL method discussed below).

The average generation values of initial good behaviour occurrence are usually highest with the linear genome method. However, the methods are based on different population sizes and individual sizes so it is difficult to draw conclusions from the raw results. With our implementation of the linear genome method (through a genome interpreter) the evolution time is similar to the time using the tree-based representation (with equivalent population size and tree size settings). The main difference between the methods is the contents of the function sets.

The ADF method uses a predefined, constant function set containing one or more ADFs. Function call acquisition occurs only through crossover with individuals of the population. The ADFs inside individuals showing proper evolved behaviour are usually quite large and complex with no noticeable patterns. It is possible that in our experiments the ADFs only provide few extra tree levels of instructions. The ADF method runs provided performance that was usually below that of the tree-based method and sometimes the worst of all HGP methods. Fig. 3.14 shows the code of a sample ADF program taken from a population with learned light avoidance behaviour.

The slowest method of function creation is the MA method. Most of the individuals in the population with proper evolved behaviour do not contain any of the functions in the module set. The creation of functions produces program size loss which in turn often lowers the entropy of the population. The behavioural performance of the MA method is usually worse than that of the tree-based method. Since similar experimental settings are used for the two methods, we can deduce that the function creation of our MA method disrupted the task learning instead of helping it. Fig. 3.15 shows the code of a sample MA program taken from a population with learned wall following behaviour.

The ARL method displays the most stable entropy and average chromosome size behaviour in most experiments. This stable behaviour is observed only with function creation, thus we think that the function creation and new individual creation processes are responsible for the stability. The method also achieves the best time and smallest deviation to reach good evolved behaviour in most experiments.

The number of functions created by the ARL algorithm depend on each run but do not grow monotonically as first expected. The function set grows and shrinks throughout the runs of the algorithm. The functions usually contain simple arithmetic operators working on function parameters. Many of the functions from populations with proper evolved behaviour contain division and addition operators that seem to calculate some form of ratio of the function parameters. Since such ratios can be helpful in all of our studied tasks, we think that some of the evolved functions are of benefit to the indi-

```
values                     defun ADF0 (ARG0, ARG1)
 -                          values
  ADF0                       -
   +                          /
    +                          *
     16                         /
     14                          9
     ADF0                        1
     12                         -
     14                          ARG1
    -                            9
     ADF0                      -
     12                         /
     9                           6
     0                           ARG1
   /                            -
    14                           ARG1
    14                           ARG0
  ADF0                        -
   +                           ARG1
    ADF0                       9
     0
     11
    13
   /
    10
    -
     13
     2
```

Fig. 3.14. Code of a sample ADF program from population showing light avoidance behaviour. ADF function definition shows characteristic large size and complex format.

viduals. Code of a sample ARL program taken from a population with learned obstacle avoidance behaviour can be found in Fig. 3.16.

Algorithms and strategies of solving problems can usually be improved to yield better solutions. Our research enables us to indicate areas of possible improvement to the studied genetic programming algorithms for the domain of robotic control. We feel that population diversity (entropy) stability, chromosome size stability, and proper fitness evaluation are the most important attributes of a well functioning genetic programming robotic controller training system. Entropy and chromosome size values should be relatively stable so that they remain at reasonable levels for a reasonable number of generations. Stability of those values depends on the definition of the fitness function and on the controller settings.

Modification of fitness function parameters leads to strong statistical and behaviour changes in the evolving population. Definition of the fitness function is thus a very important aspect for evolution of correct solutions. We choose the fitness function definitions and parameter values that produce the best performance in trial runs. However, more testing of fitness functions and their

```
values
  /
    /
      +
        Fn1056679512&6909
        Fn1056679510&11493
      /
        4
        Fn1056679514&15777
    /
      /
        /
          4
          7
        +
          s2
          4
      +
        +
          s7
          s3
        -
          s5
          0
  /
    +
      s3
      6
    -
      s2
      7
```

```
Fn1056679510&11493
  +
    s4
    s3
Fn1056679514&15777
  -
    s4
    +
      s1
      s7
Fn1056679512&6909
  *
    2
    s4
```

Fig. 3.15. Code of a sample MA program from population showing wall following behaviour.

```
values
 +
  s0
  9
 -
  +
   /
    +
     2
     s4
    +
     4
     s4
   *
    -
     -
      s5
      s3
     9
    Fn12541144&0&686
     s6
     s6
     s7
     s1
```

```
 /
  -
   /
    s7
    s1
   +
    1
    0
  +
   Fn12386896&4&862
    8
    s3
    6
    s4
   +
    6
    6
```

```
Fn12541144&0&686
 -
  +
   d0
   d1
  -
   d2
   d3
Fn12386896&4&862
 +
  -
   d0
   d1
  +
   d2
   d3
```

Fig. 3.16. Code of a sample ARL program from population showing obstacle avoidance behaviour. Characteristic ARL functions are visible.

parameters should be done to identify the optimal settings for each learning task.

The ADF method builds ADFs of initial size equivalent to the main program body. We feel that smaller building blocks (functions) are more useful for robotic controllers. The sizes of the ADFs do not decrease well enough before the population prematurely converges. We think that it would be best to specify a smaller initial and maximum size of the ADFs so that the functions require less time to find optimal configurations.

We feel that the poor performance of the MA method is due to the creation of modules which lowers the average program tree size. Since no mechanism exists to counteract this loss of program size and accompanying loss of entropy, the population often converges prematurely to suboptimal solutions. We propose that the probability-based compression and expansion operator invocation of the MA method be replaced by a need-based operator invocation (similar to that found in the ARL method). This new operator invocation should lead to better performance through adjustments of operator frequencies based on population needs.

The ARL method contains a mechanism to neutralize the bad effects of function creation. Thus, the method exhibits very stable entropy and average size behaviours while quickly evolving high performing robotic controllers. The creation of random individuals using the enriched function set at the start of a new epoch provides the genetic algorithm with fresh search material. The functions found in the adaptive representation step of the algorithm are small and seem better building blocks than the functions in the ADF method. We feel that best performance can be achieved by some kind of a dynamically evolving entropy threshold calculation.

Influx of random individuals to the population during evolution can lead to problems. Too many new random individuals can destabilize good solutions present in individuals of the previous population. We think that a low replacement fraction used with elitism of best individuals should produce the optimal evolutionary balance. Elitist individuals would always be copied into a new population and would ensure that the fittest individuals are not lost between generations.

Variation in the population can also be achieved by using a mutation operator for the tree-based representation methods. The mutation operator can quickly add subtle variety to the population. The crossover operator can perform similar mutations but with a lower probability of success based on the size and structure of the program tree.

Future work with the Khepera GP Simulator involves formulation of a proper physics model to study object interaction tasks. Modification of the simulation engine for multi-threaded robot simulations would enable proper real-time multi-robot simulation. With the use of a real Khepera robot, we hope to add serial Khepera interface to the simulator and validate the correctness of our Khepera simulation engine.

In this work, we evolve reactive, memoryless robotic controllers. Our results indicate that the controllers can be trained to exhibit some level of proper behaviour for the studied tasks. The extension to this research would be to study memory-based robotic controllers that can store previous actions and use them to decide future behaviour. Such controllers using the linear genome method have been shown in [20] to successfully and quickly evolve more complex behaviours than a memoryless controller.

We would also like to use a real Khepera robot to verify our results. Physical robots train in a noisy and sometimes unpredictable environment and would provide a real world test case for our research. Because of the reactive learning system, the simulator and robotic controllers can be easily modified to perform experiments with a real Khepera robot.

References

1. P.J. Angeline, Genetic Programming and Emergent Intelligence, In K.E. Kinnear, Jr. (ed.), Advances in Genetic Programming, chapter 4, pp. 75–98, MIT Press, 1994
2. P. J. Angeline and J. B. Pollack, The evolutionary induction of subroutines, In Proceedings of the Fourteenth Annual Conference of the Cognitive Science Society, Lawrence Erlbaum, Bloomington, Indiana, USA, 1992
3. R.C. Arkin, Behavior-Based robotics, MIT Press, Cambridge, MA, 1998
4. W. Banzhaf, P. Nordin and M. Olmer, Generating Adaptive Behavior using Function Regression within Genetic Programming and a Real Robot, In J.R. Koza, K. Deb, M. Dorigo, D.B. Fogel, M. Garzon, H. Iba, and R.L. Riolo (eds.), Genetic Programming 1997: Proceedings of the Second Annual Conference, pp. 35–43, Morgan Kaufmann, San Francisco, CA, USA, 1997
5. W. Banzhaf, D. Banscherus, and P. Dittrich, Hierarchical Genetic Programming using Local Modules, Series Computational Intelligence, Internal Report of SFB 531, No. 56/99, Univ. of Dortmund, D-44221 Dortmund, Germany, 1999
6. R. Brooks, A robust layered control system for a mobile robot, IEEE Journal of Robotics and Automation, **2**(1), 1986
7. R. Brooks, Artificial Life and Real Robots, In F. J. Varela and P. Bourgine (eds.), Toward a Practice of Autonomous Systems: Proceedings of the first European Conference on Artificial Life, pp. 3–10, MIT Press-Bradford Books, Cambridge, MA, 1992
8. K.E. Kinnear, Jr., Alternatives in Automatic Function Definition: A Comparison Of Performance, In K.E. Kinnear, Jr. (ed.), Advances in Genetic Programming, chapter 6, pp. 119–141, MIT Press, 1994
9. J.R. Koza, Genetic Programming: On the Programming of Computers by Means of Natural Selection, MIT Press, Cambridge, MA, 1992
10. J.R. Koza, Evolution of subsumption using genetic programming, In F. J. Varela and P. Bourgine (eds.), Proceedings of the First European Conference on Artificial Life, Towards a Practice of Autonomous Systems, pp. 110–119, MIT Press-Bradford Books, Cambridge, MA, 1992
11. J.R. Koza, Genetic Programming II: Automatic Discovery of Reusable Programs, MIT Press, Cambridge, MA, 1994

12. J.R. Koza and J.P. Rice, Automatic programming of robots using genetic programming, In Proceedings of Tenth National Conference on Artificial Intelligence, pp. 194–201, AAAI Press/MIT Press, 1992
13. J.R. Koza, F. H. Bennett III, D. Andre, and M. A. Keane, Genetic Programming III: Darwinian Invention and Problem Solving, Morgan Kaufmann, San Francisco, 1999
14. M. J. Mataric, Behavior-Based Control: Examples from Navigation, Learning, and Group Behavior, Journal of Experimental and Theoretical Artificial Intelligence, **9**(2–3),1997
15. P. Nordin, A Compiling Genetic Programming System that Directly Manipulates the Machine-Code, In K.E. Kinnear, Jr. (ed.), Advances in Genetic Programming, chapter 14, pp. 311–331, MIT Press, Cambridge, MA, 1994
16. P. Nordin and W. Banzhaf, Genetic Programming Controlling a Miniature Robot, In E.V. Siegel and J.R. Koza (eds.), Working Notes for the AAAI Symposium on Genetic Programming, pp. 61–67, AAAI, MIT, Cambridge, MA, USA, 1995
17. P. Nordin and W. Banzhaf, A genetic programming system learning obstacle avoiding behavior and controlling a miniature robot in real time, SysReport 4/95, University of Dortmund, Fachbereich Informatik, 1995
18. P. Nordin and W. Banzhaf, Real Time Evolution of Behavior and a World Model for a Miniature Robot using Genetic Programming, SysReport 5/95, Dept. of CS, University of Dortmund, 1995
19. P. Nordin and W. Banzhaf, An on-line method to evolve behavior and to control a miniature robot in real time with genetic programming, Adaptive Behavior, **5**(2), pp. 107–140, 1997
20. P. Nordin and W. Banzhaf, Real Time Control of a Khepera Robot using Genetic Programming, Cybernetics and Control, **26**(3), pp. 533–561, 1997
21. P. Nordin, W. Banzhaf, and M. Brameier, Evolution of a world model for a miniature robot using genetic programming, Robotics and Autonomous Systems, **25**(1–2), pp. 105–116, 1998
22. C. Reynolds, An Evolved, Vision-Based Behavioral Model of Coordinated Group Motion, In J. Meyer, H.L. Roitblat and S.W. Wilson (eds.), From Animals to Animats II: Proceedings of the Second International Conference on Simulation of Adaptive Behavior, MIT Press-Bradford Books, Cambridge, MA, 1993
23. J.P. Rosca, Entropy-Driven Adaptive Representation, In J.P. Rosca (ed.), Proceedings of the Workshop on Genetic Programming: From Theory to Real-World Applications, pp. 23–32, Tahoe City, California, USA, 1995
24. J.P. Rosca, Hierarchical Learning with Procedural Abstraction Mechanisms, PhD thesis, University of Rochester, 1997
25. J.P. Rosca and D.H. Ballard, Hierarchical Self-Organization in Genetic Programming, In Proceedings of the Eleventh International Conference on Machine Learning, Morgan Kaufmann, 1994
26. J.P. Rosca and D.H. Ballard, Discovery of Subroutines in Genetic Programming, In P.J. Angeline and K.E. Kinnear, Jr. (eds.), Advances in Genetic Programming 2, chapter 9, pp. 177–202, MIT Press, Cambridge, MA, USA, 1996
27. J.P. Rosca and D.H. Ballard, Genetic Programming with Adaptive Representations, Technical Report TR 489, University of Rochester, Computer Science Department, Rochester, NY, USA, 1994

4

Evolving Controllers for Miniature Robots

Michael Botros

Department of Computer and Electrical Engineering,
Faculty of Engineering, McMaster University,
1280 Main St. West, Hamilton, Ontario , Canada L8S 4K1,
botrosmw@mcmaster.ca

Using traditional path planning and artificial intelligence techniques has restricted the use of mobile robots to limited tasks in previously known environments, yet potential applications include dynamic and unstructured environments. One of the very promising methods of designing controllers for autonomous and mobile robots is using Evolutionary Computations, a class of algorithms which mimics the natural evolution process.

In this chapter we present a series of experiments in evolutionary robotics that used the miniature mobile robot Khepera. Khepera robot is widely used in evolutionary experiments due to its small size and light weight which simplify the setup of the environments needed for the experiments. The controllers evolved by the presented experiments include classical and spiking neural networks controllers, fuzzy logic controllers and computer program obtained by Genetic Programming. The tasks performed by the robots through the experiments reflect learning many basic as well as high level behaviors. These behaviors include: navigating in dynamic environment with static or dynamic obstacles, seeking and following the light sources present in the environment, returning home for recharging the battery, and collecting trash objects from the environment. The chapter also presents an experiment in co-evolution in which a predator-prey behavior is learned by two robots. The chapter ends with an experiment that evolves spiking neural networks, a new artificial neural networks model that accurately models the biological neuron activation. This experiment presents the use of evolution to obtain a spiking neural network that enables the robot to navigate depending only on vision information.

4.1 Introduction

Khepera is a miniature mobile robot that is widely used in laboratories and universities in conducting experiments aiming at developing new control algorithms for autonomous robots. It was developed by the Swiss Federal Institute of Technology and manufactured by K-team [1] [2]. Khepera robot is cylindrical in shape with a diameter of 55 mm and a height of 30 mm. Its weight is about 70 gm. Its small size and weight made it ideal robotic platform for experiments of control algorithms that could be carried out in small environments such as a desktop.

The robot is supported by two wheels; each wheel is controlled by a DC motor that can rotate in both directions. The variation of the velocities of the two wheels, magnitude and direction, will result in wide variety of resulting trajectories. For example if the two wheels rotate with equal speeds and in same direction, the robot will move in straight line, but if the two velocities are equal in magnitude but different in direction the robot will rotate around its axis.

Fig. 4.1. Miniature mobile robot Khepera (with permission of K-team).

The robot is equipped with eight infrared sensors. Six of the sensors are distributed on the front side of the robot while the other two are placed on its back. The exact position of the sensors is shown in figure (4.2). The same sensor hardware can act as both ambient light intensity sensor and proximity sensor.

Each of the eight sensors consists of emitter and receiver parts so that these sensors can function as proximity sensors or ambient light sensors. To function as proximity sensors, it emits light and receive the reflected light intensity. The measured value is the difference between the received light intensity and the ambient light. This reading has range [0, 1023] and it gives a rough estimate how far the obstacles are. The higher reflected light intensity the closer obstacles are. It should be noted that we cannot find a direct mapping between the sensor reading and the distance from the obstacle, as this reading depends on factors other than the distance to the obstacle such as the color of the obstacle.

To function as ambient light sensors, sensors use only receiver part of the device to measure the ambient light intensity and return a value that falls in

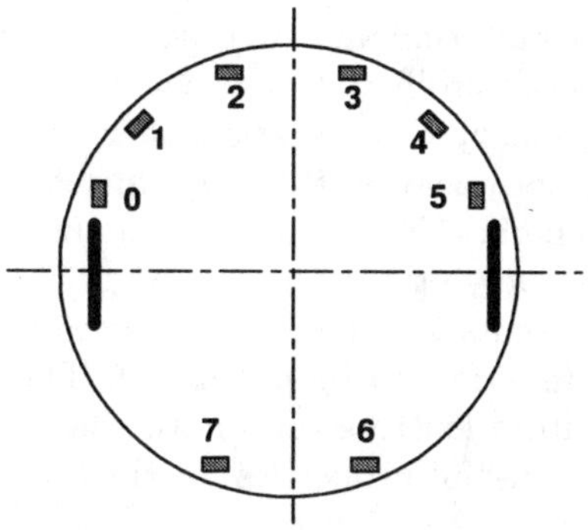

Fig. 4.2. The position of the eight sensors on the robot (with permission of K-team)

the range of [0, 1023]. Again, these measurements depend very strongly on many factors such as the distance to the light source and its direction.

An interesting feature of the Khepera robot is its autonomy, which includes autonomy of power and control algorithm. For the purpose of power autonomy, the robot is equipped with rechargeable batteries that can last for about 45 minutes. For experiments that may require much longer time, the robot can be connected to a host computer by a lightweight cable to provide it with the needed electrical power. This is an important feature that allowed long control experiments (such as developing evolutionary algorithms) to be carried out without repetitive recharging.

On the other hand, for the control autonomy, the robot's CPU board is equipped with MC68331 microcontroller with 512K bytes of ROM (system memory) and 256K bytes of RAM (user memory). This RAM memory can accommodate reasonable length program codes to provide control autonomy. The robot can be programmed using Cross-C compiler and the program will be uploaded to the robot through serial port communication with a host computer. Also the robot can be remotely controlled by a host computer where the control commands are sent to the robot through the serial link connection mentioned above. This mode of operation has an advantage of using computational power of the host computer.

4.2 Evolutionary Computations and Robotics

The term Evolutionary Computation is used to describe a set of algorithms that use the idea of evolution in solving complex computational problems such as our problem of designing a robot controller. It includes algorithms such as Genetic Algorithms GA, Genetic Programming GP and Evolutionary Strategies. They operate on a population or a group of individuals each representing a proposed solution of the problem. Then they apply a set of biologically inspired operators such as mutation and crossover to obtain a better generation which is more suited to the problem to be solved.

So what can Evolutionary Computation offer to robotics? First thing it offers to robotics is an optimization tool. Optimization is a frequent type of problems solved by Genetic Algorithms due to the embedded competition between individuals. In applying the Genetic Algorithm for optimization, the individuals are usually points in the space to be searched for optimum point and the fitness is the function to be optimized. The reproduction aims at generating new points from existing ones until the optimum point is found. Genetic Algorithm offers useful properties for the optimization problem:

- It is applicable to continuous, discrete and mixed optimization problems and it requires no information about the continuity or the differentiability of the function to be optimized. It also can be used for problems of optimization with constraints. The constraints on the parameters to be optimized can be easily translated to constraints on the genetic operators to produce individuals inside the search domain defined by the constraints.
- Genetic Algorithms are suitable for many practical problems that require multi-objective functions. Multi-objective optimization can be accomplished by designing fitness function that is a weighted sum of required objectives. Another solution is using Co-evolution where multiple populations are used instead of single population. Each population is bred to optimize certain objective while individuals are exchanged between them (migration).

For example, one of the possible methods for evolving a neural network controller is to let the evolutionary algorithm choose the optimal weights of the neural network, so the problem of evolving this controller to perform obstacle avoidance behavior can be viewed as a problem of optimizing the different weights. Also this problem is multi-objective optimization because we want the neural network to achieve different goals such as avoiding the obstacles while keeping a reasonable velocity and keeping a straight path.

Second thing evolutionary computations can offer to robotics is providing a method of learning rules necessary for the robot to achieve some task. In this case the controller is mainly a set of rules and we want to choose the optimal set of rules that serve this task. Programming the rules by hand or testing different combinations of them is a tedious task. An example of using the genetic algorithm to learn robots rules is a system built at the Naval Research Laboratories and is called SAMUEL [3]. It used the method described above to learn Nomad robot navigation and obstacle avoidance. Rules are not the only form of controllers that can be designed by evolutionary computations [4] [5]. In the next sections see how evolutionary computations can be used to design controllers such as neural networks and fuzzy logic controllers.

4.3 Evolving Neural Network Controllers

Many researchers have found neural network and interesting solution for the problem of the building behaviors for the Khepera robot. The ability to learn and the ability to deal with noisy sensors were apparent advantages in favor of the neural network.

Different approaches exist for designing neural network controllers. One approach is to use neural networks learning algorithms to train the synaptic weights. Example of this work can be found in [6]. Another Approach is to use the Genetic Algorithm as a search or optimization tool to find the best neural network controller through the evolution process. The leading work of evolving neural network controller for a real Khepera robot was done by Floreano and Mondada [7]. They evolved a simple feed forward neural network that consisted of input and output layers with no hidden layers. The neural network controller enabled the robot to navigate in the arena while avoiding obstacles .

We can use the genetic algorithm in different ways to evolve neural networks. It can be used to search for the optimal synaptic weights, or to search for the optimal network architecture along with the synaptic weights. Also, it can be used to evolve the learning parameters needed to train the neural network. Examples of these methods are presented in the following subsections. For example, using the genetic algorithm to search for the suitable synaptic weights given a predefined architecture is presented in experiments 1 and 3 whose goals are to evolve obstacle avoidance and home seeking behaviors respectively. On the other hand, evolving the network architecture is the method used in experiment 2 to develop a light seeking behavior. Finally, evolving Hebbian learning rules and the rate of learning is an example of evolving the learning parameters of the neural network and it is one of the methods used in experiment 5 to co-evolve predator-prey behavior in two robots.

4.3.1 Experiment 1: Evolving Obstacle Avoidance Behavior

The goal of this experiment [8] is to evolve a neural network controller for obstacle avoidance navigation in environments with static or dynamic obstacles. The proposed neural network is a feed forward neural network with input layer consisting of 8 neurons, hidden layers of 2 neurons and output layer of 2 other neurons. The inputs of the neural network are the eight proximity sensors that are arranged on the robot as shown in figure (4.2). The input range of each sensor is [0, 1023]. The values of the inputs were scaled to the range [0, 1] before being applied to the neural network. The Outputs of the neural network controller are applied to the motors of left and right wheels. The activation function of the neurons is the sigmoid function which is limited between [-1, 1], so the output of the neural network had to be properly scaled before being applied to the motors.

The fitness function used rewarded the individual which moves with a suitable forward speed and penalize the individual which rotates around itself or comes close to the obstacle. It has the following formula:

$$\text{fitness} = C_1(V_L + V_R - |V_L - V_R|) - C_2 \sum_{i=1}^{8} S_i \tag{4.1}$$

where V_L, V_R are the velocities of left motor, right motor respectively, S_i is the proximity sensor number i, and C_1, C_2 are suitable positive scaling factors. The term $V_L + V_R$ will maximize the forward speed while term $|V_L - V_R|$ will minimize the rotation of the robot which occurs due the difference between the velocities of left and right wheels. Also, the robot will learn to keep a suitable distance separating it from the obstacles in order to decrease the magnitude of the sum of the sensors. The constants C_1, C_2 set the relative importance of each component of the fitness function, for example increasing C_2 will emphasize the importance of avoiding obstacles relative to keeping a straight path.

The fitness of the individuals is evaluated as follows: each individual was allowed to perform a 400 time step, in each step it reads the proximity sensors, calculate the output speeds using its own neural network and apply these speeds to the motors then it measures the new proximity sensor values and calculate its fitness function according to the above formula. Individual fitness is the sum of its fitness function over the 400 time steps. The above algorithm lasted for 120 generations.

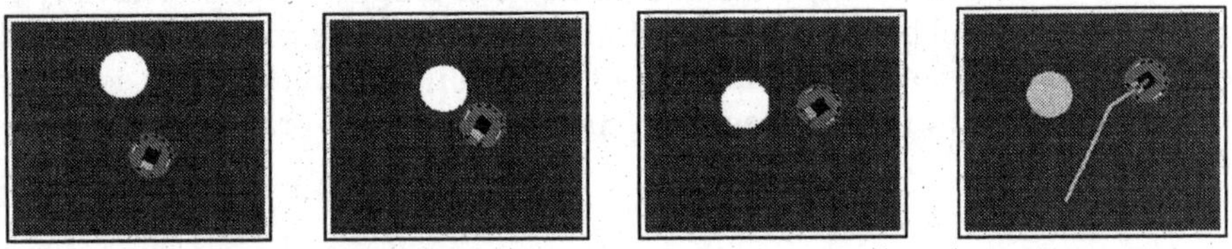

Fig. 4.3. Trajectory of the robot in an environment with moving obstacle.

The result of the experiment showed successful emergence of the desired behavior. After 80 generations, the robot was able to move in straight trajectories and it learned to keep a suitable distance between its path and the obstacles or walls. This is clear in the left section of figure (4.4) which shows the behavior of the robot in an environment with large centered obstacle. While moving parallel to the wall, the robot moves in a straight path and maintains certain distance between its path and the wall. Fig. (4.3) shows the behavior of the best fit individual when a round object of the same size of the robot is approaching its path. The slides taken from the motion of the robot shows its turning and avoiding collision with the moving object. Fig. (4.4) shows the behavior of the robot in an environment with obstacles having

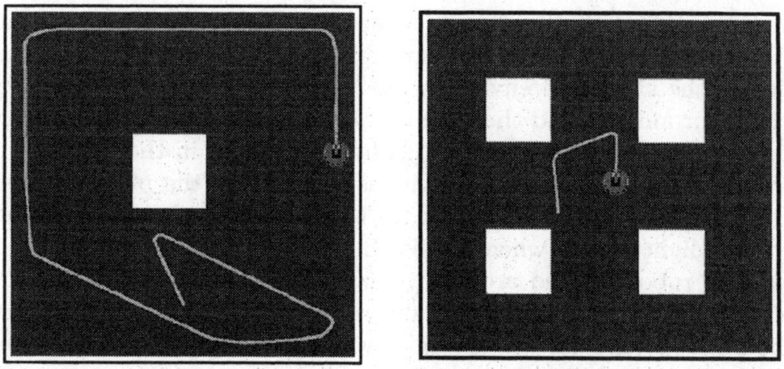

Fig. 4.4. Trajectories of the robot in environments with large obstacles with sharp corners .

sharp corners which is difficult to detect if the robot is heading towards the corner. We can see that the robot turns before being close to the corner and this behavior is repeated twice. It should be also noted in this environment that distance between the two obstacles is about twice the diameter of the robot.

4.3.2 Experiment 2: Evolving Light Seeking Behavior

This experiment was performed by Hülse et al. [9]. The goal of the experiment is to evolve a neural network controller that enables the robot to seek the light source available in its arena. The proposed neural network had 16 input neurons and 2 output neurons. The input neurons corresponds to the 8 proximity sensors and the 8 ambient light sensors while the two output neurons correspond to the two motor speeds.

The evolutionary algorithm used in this experiment allowed the evolution of the structure of the neural network along with the synaptic weights values. It can evolve the number of the hidden neurons necessary to connect the input and output layers along with their recurrent connections.

The evolution experiment was carried in a simulated environment while the best fit individual was tested in both real and simulated environments. The results of the experiments showed the emergence of light seeking behavior in the early generations. The best fit individual was tested in two simulated environments and in a physical environment. The first simulated environment contained one light source. The robot was able to move towards the light source from different starting positions. The second simulated environment contained more than one light source. The robot moved towards the nearest light source. The best fit controller was then moved to a real robot and tested in a physical environment. In similar conditions to the simulated environment, the robot was able to move to the light source. The environment was slightly modified to test the controller ability to adapt to changes in the physical environment. When the light source was moved the robot was still able to move towards and follow the light source, which shows consistency with the behavior in the simulated environment. Next, the light source was removed from the environment, and then the robot started to move in curved or semi circular trajectories compared to straight trajectories in the presence of the light source. To test how the behavior is affected by the proximity sensors, the proximity sensors were removed, in this case the robot was still able to move to the light source when it existed in the environment, however in its absence, the robot rotated around its axis. These results show good match between the behavior in simulated and real environments, they also showed that the evolved behavior was invariant when the light source was moved but was affected by removing the connections from the proximity sensors when there was no light source in the environment [9].

We notice in this experiment that the genetic algorithm allowed the evolution of the network architecture along with the best synaptic weights . This method enables the genetic algorithm to search for the best neural network controller in the space of the network architectures. In general, this method would lead to better quality solution than the case of predefined network architecture. On the other hand, this method requires a variable length chromosome that encodes the neural network. Also the chromosome is expected to be longer than the one that encodes only the synaptic weight which would result in longer evolution time.

4.3.3 Experiment 3: Evolving Recharging and Home Seeking Behavior

This experiment was performed by Floreano and Mondada [10]. Although the experiment evolved an interesting home seeking behavior, the actual goal of the experiment was to show that behaviors can be evolved without being explicitly included in the fitness function. In this experiment the fitness function didn't include a pleasure part to reward the robots when returning to home (or the recharging area). However, without recharging, the robot will not be able to live longer and achieve a high fitness which was allowed to be calculated over a period longer than the battery life time.

The experiment was conducted in a rectangular environment where one of the corners was illuminated with a tower carrying a number of lamps. This corner was considered the robot's home or recharging area. In this corner, a circular sector of the ground is painted in black such that the robot can detect it using an extra ambient light sensor placed under the robot. This sensor is active in the entire environment except the recharging area.

Using the robot actual battery which lasts for 40-45 minutes will cause the experiment to last for a very long time. Instead, the robot was equipped with a simulated battery that discharges linearly with time in a maximum of 20 seconds. The reading of the battery time can be considered a virtual battery sensor whose value falls between [0, 1], with 1 indicating that the battery is fully charged. For the robot to detect the light source associated with its recharging area, two sensors acted as ambient light sensors beside their function as proximity sensors. The two sensors are the ones labeled 2 and 6 in figure (4.2).

The neural network controller used was 3 layers neural network with recurrent connections in the hidden layer The input layer has 8 neurons for proximity sensors, 2 neurons for ambient light sensors and 2 other neurons for floor brightness and simulated battery sensor. The output layer consisted of 2 neurons that correspond to the motor speeds.

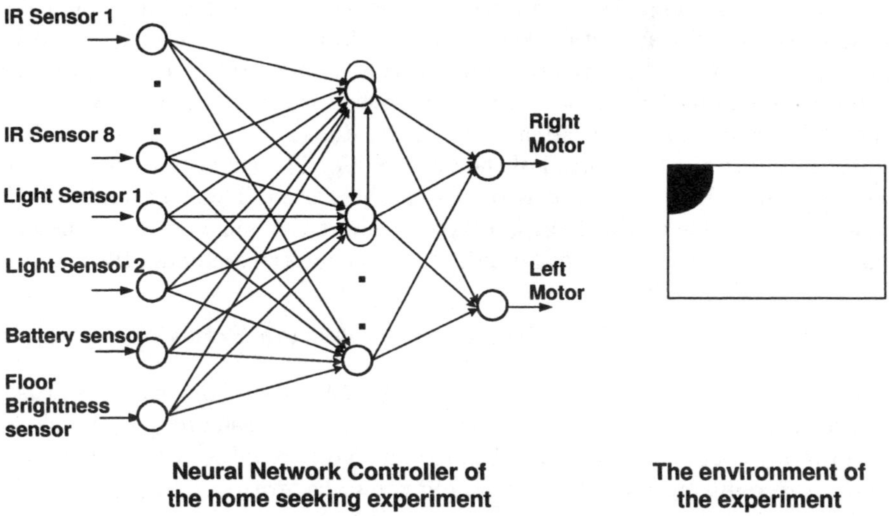

Fig. 4.5. The neural network controller of the home seeking experiment (left). A figure of the environment(right).

The fitness function used in the experiment rewarded the individuals that move with large speed and avoid the walls. The fitness function formula is given by [10]:

$$\text{fitness} = v(1 - i) \tag{4.2}$$

where v is normalized average speed of the two motors $0 \leq v \leq 1$, and i is normalized value of the maximum proximity sensor $0 \leq i \leq 1$. The fitness function is calculated and summed over maximum number of 150 time step while the battery life lasts for 20 seconds or 50 time steps. Also the fitness function is not summed when the robot is in the recharging area. The robot should learn to return to the recharging area before its battery life comes to an end. Furthermore, it should not stay there for long since no fitness is gained there. This behavior is not stated explicitly in the fitness function but implicitly implied by the conditions of the experiment.

The genetic algorithm lasted for 240 generations. The results of the experiment showed that in the last generations the behavior of the robot was as expected. It returned to home for recharging without spending much time there after recharging. The behavior of the best fit individuals was as follows: When it was placed in the charging area, it quickly moved away and returned only before the the battery life ends by 5 time steps. Outside the recharging area, it moved with maximum speed avoiding the walls whenever they are encountered. Testing the best fit individuals from different initial positions showed that it was able to return for recharging for many times for most of the initial positions.

Also the results of the experiments showed that we can find a direct relation between the activation level of one of hidden neurons and certain behaviors. Observing the activation level of this hidden node over the robot life showed that it had a low activation level when the robot navigated outside the recharging area but gradually increased during the journey to the back for charging in the last period of the battery life. The activation level reached its maximum when the robot is in the charging area. This fact supports the assertion that this hidden neuron played a role in the behavior responsible for planning the journey back to home before the battery life ends [10].

4.3.4 Experiment 4: Evolving Trash Collection Behavior

This experiment was performed by Nolfi [11]. The goal of the experiment is to teach the Khepera robot how to clear the arena from trash objects by grasping and placing them near the walls of the arena. This complex task requires skills such as recognizing the trash object and the walls, grasping and releasing the object, and obstacle avoidance. To accomplish this task the Khepera robot is provided with a gripper module that is added on the top of the robot (see figure 4.6). The gripper can perform two main actions: picking and releasing the object. The robot can detect the presence of an object in the gripper by using a light barrier sensor placed in the gripper.

One approach to teach the robot this complex task is to split it into a set of simpler tasks or behaviors and design a module that control each behavior then designing a coordination method that decides which of these modules will take control of the robot based on the current situation. Each behavior could be designed by hand, evolved or learned by other learning methods. An example of this approach is found in [12] where all the modules are programmed by hand except the grasping behavior which was learned using reinforcement learning. However, in the experiment that we will present the goal was to evolve the entire behavior and to test the hypothesis that different modules of the evolved neural network correspond to certain basic behaviors.

The experiment evolved five different neural network architectures among them two with modular structure. All the architectures had 7 input neurons and 4 output neurons. The input neurons correspond to the 6 proximity sensors on the front side of the robot and the barrier light sensor present in the

Fig. 4.6. Khepera robot with the additional gripper module (with permission of K-team).

gripper. The output neurons are the 2 motor speeds and the 2 actions of the gripper. The five neural network architectures had the following structures:

1. The first neural network is a feed forward neural network with no hidden layer.
2. The second neural network is also a feed forward neural network but with a hidden layer of 4 neurons.
3. The third neural network has recurrent connections between two extra input and output nodes.
4. The fourth neural network has a modular structure. It has two modules each with its own set of the four output neurons. Each module takes control in different predefined situations. The first module takes control when the robot is looking for the trash object and grasping it. Its goal is recognizing the trash object. The second module takes control when the robot is holding the trash object and heading towards the wall. Its goal is recognizing the wall and avoiding obstacles while holding the trash object.
5. The fifth neural network has modular structure too. It consists of two modules. Each module has its own four output neurons in addition to four selector neurons. The selector neurons compete with each other to decide which module will take the control. For example, if at a certain time the activation level of the selector neuron of the left motor is higher in the first module, then output of neuron corresponding to the left motor in the first module will be sent to the left motor.

The environment used in the evolution process was an arena with walls of height 3 cm and it contained 5 trash objects which are cylindrical in shape.

The genetic algorithm used population of 100 individuals for each of the five architectures and it lasted for 1000 generations. The fitness function essentially rewarded individuals for the number of the trash objects successfully placed outside the arena with less rewards for objects that the robot was only successful to pick. Each individual was tested for 15 epochs and its fitness valuation was the sum of its fitness function in each epoch.

The experiment described above was repeated 10 times for every architecture. The 10 best individuals of each architecture were given the same task of clearing the arena from 5 trash objects. The results showed that the fifth neural network excelled the others where 7 of its best 10 individuals were able to successfully complete the task. Only one or two individuals were able to complete the task for the other architectures.

Considering the hypothesis that modular architecture may contain modules that correspond to certain behavior, it was found that the best individual of the fifth architecture use both modules for controlling the left motor and uses only one module for rest of the outputs. This fact showed that relation between modules and basic behaviors could not be proven in this experiment [11]. However, in the experiment of home seeking and battery recharging certain hidden neuron was shown to be responsible for detecting low battery and returning home for recharging.

4.3.5 Experiment 5: Co-evolving Predator-Prey Behavior

By co-evolution we mean evolving two competing populations simultaneously such that the fitness evaluation of one is at the expense of the other. The co-evolution adds more competition stress to the evolution process which is, by nature, characterized by the competition for survival among individuals of the same generation. We are now going to present an interesting experiment in co-evolution whose goal was evolving a predatory-prey behavior in two khepera robots. The predator robot is required to chase the prey robot and contact it.

The experiment was performed by Floreano and Nolfi [13] [14]. In the experiment, the predator robot is equipped with a vision module (see figure 4.7) to recognize the prey robot which was provided with a black perturbation that can be easily detected on the white walls of the environment. To provide fair competition, the maximum speed of the prey robot is allowed to be twice that of the predator robot.

The environment was a square one of dimension 47 cm. That size was chosen such that prey will always be within the detection range of the vision module of the predator which can detect objects in range of 5 to 50 cm. The evolution experiment was carried in a simulated environment of the same details of the actual one. This will help to decrease the time of the evolution and to avoid the hardware problems resulting from the twisting of the power cables of the two robots.

The K213 vision module of the khepera robot is an additional module that is connected to the top of the robot. It is cable of providing a linear image of 64 pixels that cover a vision angle of 36 degrees. Furthermore, the module has a microcontroller that can process the image data and instead of sending the 64 bytes of the image to the robot it can detect the least eight pixels in intensity and pass them to the robot.

Fig. 4.7. Khepera robot with the extra K213 vision module (with permission of K-team).

In the simulated computer environment, the experiment designers divided the vision range to 5 sections each representing a simulated photosensor. These simulated photosensors act as input for the neural network controller of the predator robot. A simulated photosensor is considered active if a pixel of minimal intensity is within its range, possibly because of the presence of the prey robot in this section.

The controllers of the two robots are shown in figure (4.8). Each controller is recurrent neural network. The predator neural network has extra 5 input neurons corresponding to the five photosensors. On the other hand, the two outputs of the prey neural network are multiplied by a factor of two before being applied to the motors of the robot.

The genetic algorithm used two competitive populations each of 100 individuals and the experiment lasted for 100 generations. As we mentioned earlier, the fitness evaluation of each robot is at the expense of the other. The predator robot is awarded for decreasing the time needed to contact the prey. Its fitness is a normalized version of that time and falls in the range of [0, 1]. The prey robot fitness function is just (1 − predator fitness). The fitness function of each individual, predator or prey, is evaluated through testing it against the best individuals of the last 10 generations of the opposite type.

The experiment used direct encoding to encode the synaptic weights of the neural network. Each weight is encoded in 5 bits. The first bit is always used to encode the sign while the other four bits differed according to the instance of the experiment. We will to summarize each of the three instances of the experiments along with its results [13] [14].

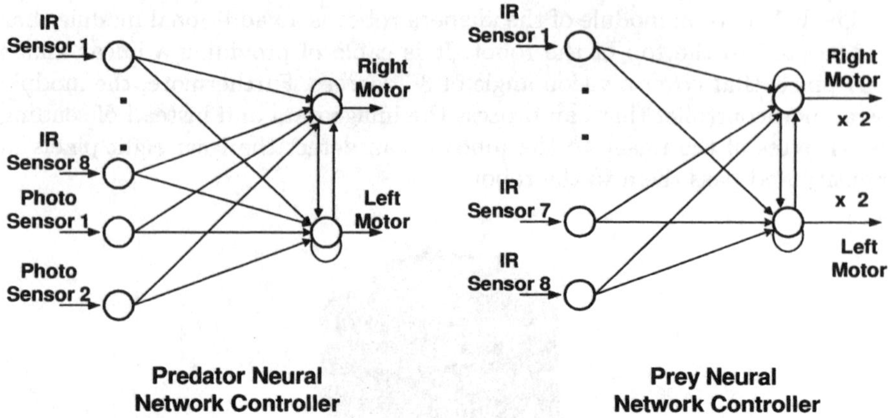

Fig. 4.8. The neural network controller of the predator and prey robots.

- First instance of the experiment: In this instance the four bits simply encoded the value of the synaptic weight which falls in the range of [0, 1]. The results of this instance of the experiment showed that there was no population superior to the other all the time span of the evolution. In the first generations the predator was able to chase the prey and contact it. After 70 generations, the prey was cable of turning away when the predator approached it. After 90 generations, the predator learned better attacking methods for chasing the prey.
- Second instance of the experiment: Only two bits were used to encode the value of the weight and the other two bits are used to encode four different level of uniform noise that would be added to the weights. The results of this instance of the experiment showed that the noise level in the synaptic weights of the prey was higher than those of the predator which suggested that the prey made use of this noise to evolve an unpredictable and changing trajectory to confuse the predator robot.
- Third instance of the experiment: The four bits are used to encode the learning parameters of the synaptic weights rather than the value of the weights. Two bits encoded the Hebbian rules and the other two bits encoded the learning rate. The value of the weight is randomly generated between [0, 1] and continuously updated according to the rules. The results of this instance of the experiment showed that the average fitness of the predator is higher than that of the prey. In terms of the apparent behavior, it developed better chasing techniques than that of the first instance of the experiment. In terms of the synaptic weights, the experiment results showed that the synaptic weights were adjusted by the Hebbian learning and the resulting motor speed steered the robot towards the prey, a property which require fine tuning of the weight values if the encoding method of the first instance was used.

The results of this experiment are interesting and reflect how the behavior of the robot was dependent on the types of the parameters of the controller encoded in the gene despite the fact that the controller had the same architecture in the three instances of the experiment. We would expect also that different behaviors could have obtained by allowing the evolution of the architecture of the neural network along with the weights.

4.4 Evolving Fuzzy Logic Controllers

Fuzzy Logic is a mathematical tool that can manipulate human vague concepts and linguistic variables. Zadeh in [15] proposed a method to treat human knowledge based on the Theory of Approximate Reasoning. He proposed that systems with ill defined or with uncertain model can be treated by fuzzy logic. These principles were then used to build a controller for the first time in [16].

In this section, we will briefly present how the fuzzy controller can be applied to the problem of mobile robot navigation and obstacle avoidance. The fuzzy controller usually consists of three parts:

The Fuzzifier

The first step in any fuzzy control application is to specify the fuzzy sets and the corresponding membership functions for each of input or output variables. This process is known as fuzzification. If we apply this to the Khepera input proximity sensor values, we will find that each sensor has a reading value in the range [0,1023]. One of the proposed methods for fuzzification could be: " Near","Medium", and "Far". Also membership function can have other shapes such as the triangular shape or bell shaped. See figure (4.9).

In our example of the Khepera proximity sensor, the reading 300 may have a membership in the fuzzy set "Near" that is equal to 0.75 while the membership in the sets "Medium" and "Far" are equal to 0.25 and 0 respectively. It is clear here that the crisp value 300 has been assigned a membership value for every fuzzy set defined over the range [0, 1023]. Also the output variables (left motor speed and right motor speed) can be fuzzified in the same sense. The fuzzy sets could be “Positive Large”, “Positive” , “Zero”, “Negative”, “Negative Large”.

The Fuzzy Rules

This is the main part of the controller where human knowledge can be represented in the form of if-then rules. The rule usually takes the following form:

If (antecedent part) *then* (consequent part)

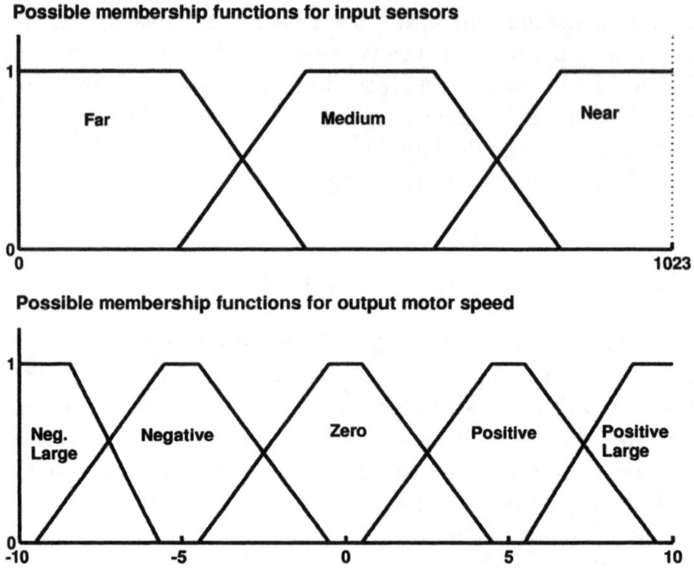

Fig. 4.9. Possible membership functions for input sensor and output motor speed.

Where the antecedent part checks the input variables and the consequent part sets one or more of the output variables. For our case of Khepera robot navigation, one of the rules can be:

If (left proximity sensor is "Near") *then*
(left speed is "Positive Large") and (right speed is "Postive")

This rule tells the robot to turn to right (by moving the left wheel faster than the right wheel) if obstacle is found on the left of the robot. If the left proximity sensor is near with membership value 0.75, then this rule will have firing value equals to 0.75. A group of fuzzy rules resembling the previous one are needed for the safe navigation of the robot.

The Defuzzifier

The outputs (left and right speeds in our case) need to be crisp values, this will be the role of the defuzzifier to convert them form fuzzy sets to crisp value. This is done through the fusion of different rules based on their firing values.

Since the performance of the fuzzy logic controllers depends on the parameters of the membership functions and the rules used, then we need to search for the best membership functions and the optimal set of rules. This leads us to thinking of genetic algorithm to evolve the best fuzzy logic controller parameters instead of designing it based on the human experience.

4.4.1 Experiment 6: Evolving Corridor Following Behavior

This experiment was performed by Lee and Cho [17]. The goal of the experiment was to evolve a fuzzy logic controller that can enable the robot to avoid the obstacles and follow the corridors of the environment. The fuzzy logic controller had 8 inputs corresponding to the 8 proximity sensors of the robots and 2 output neurons that correspond to the motor speeds. The role of the genetic algorithm in designing the controller was to evolve the best membership functions of the inputs and the outputs along with the necessary rules.

The experiment designers chose to divide the input sensory range [0, 1023] into four triangular membership functions. The same number and type of the membership functions were used for the outputs. The parameters of these functions, such as their starting and ending point on the input or output range, were binary encoded in the chromosome . Also the chromosome included information about a set of 10 possible rules.

To encourage the robot to explore the arena and follow the corridors without colliding with their walls, the fitness function had a positive part that is function of the total distance moved and the number of the check points in the arena that the robot passed through. It also has negative part that is function of the number of collisions.

The results of the experiment showed that the best fit individual was able to develop basic behaviors of avoiding collision and following walls. The performance of this evolved fuzzy logic controller was tested in two other simulated environments in which it was observed that the robot developed three distinct sub-behaviors which are: passing corridors, wall following and obstacle avoiding. The corridor passing behavior is active when the robot is moving in a narrow path with obstacles on both sides. The wall following behavior become active when the obstacles or walls are sensed on one side of the robot while the obstacle avoidance behavior become active when obstacles are sensed in front of the robot. A relation could be found between each sub behavior and a subset of the fuzzy rules that support this sub behavior. The robot switched from one sub behavior to the other depending on the current situation till its target was reached [17].

4.5 Evolving Controlling Programs

Genetic programming GP applies the evolution model to computer programs. The individuals here are computer programs that represent potential solution to required problem. Usually these problems are too complex or time consuming to be programmed by hand. An example of this type of problems is writing a program to control a mobile robot to navigate and avoid obstacles in a new environment.

Now the question that may arise is how to represent computer programs as individuals and how to design genetic operators, such as crossover and mutation, that is applicable to computer programs. Answers of these questions are in Koza's suggestion [18] of representing programs as trees that is composed of nodes and branches. The nodes are the operators that can take any value from certain function set such as {multiplication, addition..}. The branches are the operands which can be constants, input values or results of another node. Fig. (4.10) shows an example of a tree that represents a simple program.

This tree representation provided a method for performing crossover between two individuals. This is preformed by exchanging parts of the two trees representing the two individuals. To perform mutation operator we need to make sure that the resulting individuals represents a valid computer program. For example the mutation operator can take place by changing the operator in the node by another operator from the function set or by mutating the constants in the operands.

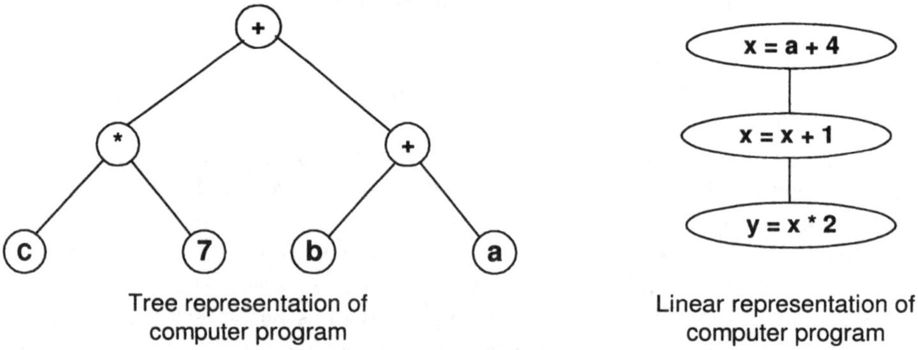

Fig. 4.10. Tree representation of computer programs versus linear representation

Having this brief overview of the Genetic Programming GP, we are now ready to present the following experiment in evolving obstacle avoidance controller program using Genetic Programming.

4.5.1 Experiment 7: Evolving Obstacle Avoidance behavior using Genetic Programming

This experiment was performed by Nordin and Banzhaf [19]. The goal of the experiment was to evolve a controller program for obstacle avoidance navigation using genetic programming. The experiment was carried on a real khepera robot in two different environments. The first environment was a rectangular arena of size 30 x 40 cm with regular walls while the other is larger in size with obstacles in its center and characterized by irregular walls. In both cases, the khepera robot was controlled by a computer workstation through a serial cable.

Motivated by applying genetic programming on real robots and obtaining a reasonable behavior in a short time, the experiment designers made two choices. First choice was not to use the tree structure we discussed above. Instead, the individual programs were represented as a linear sequence of operations along with their operands.An example of this representation is showed in figure (4.10). Second choice was to represent these instructions in the low level binary format of the controlling workstation (Sun 4). Using this representation, the crossover operators will be carried by exchanging two segments of instructions between two individual programs. The mutation operator was restricted to produce only valid machine instructions.

The population size of the experiment was small and consisted of 50 individuals and tournament selection is used when individuals are needed to be selected for crossover or mutation. The tournament works as follows: First we select n individuals from the population size N and each of the n individuals is tested and its fitness is evaluated, then we choose the best fit individual out of them for crossover and mutation.

The results of the experiment showed successful evolution of the obstacle avoidance behavior in both of the environments. In the first environment, it took the robot 20 minutes to evolve a reasonable obstacle avoidance behavior. In the second environment, it took the robot some longer time compared to learn the same behavior. This may be because of the complexity of the second environment [19].

The results of this experiment showed how the choice of some parameters of the genetic algorithm such as the encoding and selection methods, in addition to the machine format of the programs, helped in evolving the required behavior in small amount of time. We could see that a reasonable behavior emerged in less than an hour in both environments.

4.6 Evolving Spiking Neural Network Controllers

In this section, we are going to introduce a new model of the biological neurons that models the dynamical nature of neurons communication. This new model is what we call spiking neurons. We will also present an evolution experiment that evolved spiking neural network for controlling a robot based on vision information only.

To explain the spiking neuron model, we will need first to have a look at the actual way of communication between biological neurons. Biological neurons communicate by sending a large number of short pulses each second. These short pulses are known as spikes. The classical model of neurons considers only the rate of these spikes. The current activation level in the classical model corresponds to the current rate of spikes normalized by its maximum value. On the other hand, the spiking neuron provides more complex model of neuron activation function that depends on the timing between spikes.

One widely used model of spiking neuron is the "Integrate and Fire" model. In this model, the activation of the neuron is described by its membrane potential. Each spike received contributes to the membrane potential according to two factors: the weight of its synaptic connection and the time elapsed since its firing. When the accumulated effect of these spikes cause the membrane potential to go above certain threshold, the neuron fires a spike. After firing the spike, the neuron becomes unable to fire another spike instantaneously. It needs a refractory period η before it sends another spike. This refractory time depends on a certain time constant τ_m of the membrane.

At any time t, the effect of a spike on the neuron potential is a function of the time difference between the current time t and the firing time of the spike t_{firing}. This function $\epsilon(t - t_{firing})$ can be modeled by a pulse shaped function as shown in figure (4.11). In the figure, the period Δ of zero effect corresponds to the time required by spike to reach the neuron. One of the suggested expressions for $\epsilon(t - t_{firing})$ is given by [20], [21]:

$$\epsilon(s) = \begin{cases} \exp(-\frac{s-\Delta}{\tau_m})(1 - \exp(-\frac{s-\Delta}{\tau_s})) & s \geq \Delta \\ 0 & s < \Delta \end{cases} \tag{4.3}$$

where $s = t - t_{firing}$ represents the time elapsed since the firing of the spike, τ_s is the synapse time constant. Also we can model the refractory period $\eta(s)$ by a negative decaying exponential where the potential of the neuron is set after emitting the spike to a very low negative voltage to prevent emitting another spike immediately. One of the suggested expressions is given by [20], [21]:

$$\eta(s) = -\exp(-\frac{s}{\tau_m}) \tag{4.4}$$

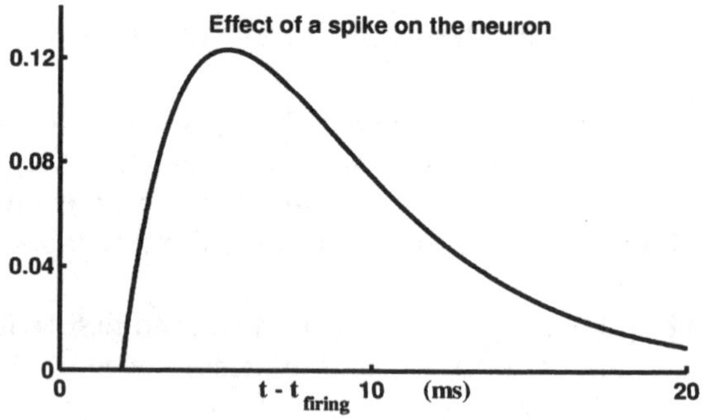

Fig. 4.11. The effect of a spike on the neuron $\epsilon(s)$

Now, we can write the the mathematical model of the spiking neuron the gives the potential of neuron i as result of addition of to quantities. The first is due to the effect of received spikes and can be written as the sum

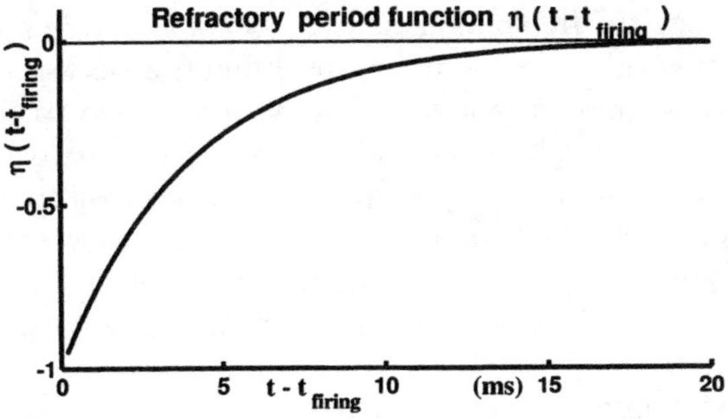

Fig. 4.12. Refractory period function $\eta(s)$

of the incoming spikes $\epsilon_j(s_j)$ from other neurons, labeled by index j, with each spike effect multiplied by the weight of its synaptic connection w_j^t. The second quantity is due to the spikes emitted by neuron i itself and can be written as sum of all refractory functions resulting form the emitted spikes. A mathematical formula of what we have just described can be given by [22]:

$$v_i(t) = \sum_j w_j^t \sum_{All\ recieved\ spikes} \epsilon_j(s_j) + \sum_{All\ emitted\ spikes} (s_i) \qquad (4.5)$$

The above equation describes the model of the activation of the neuron, represented by its membrane voltage, which takes into the consideration the timing of the emitted and received spikes in contributing to the membrane potential. A question might arise here asking why we would be interested in more complex model for neural network to employ is robot controllers. The answer is that model should be better at detecting the time varying relation between the sensors and motors due to its dynamic nature [22]. In the rest of this section, we will see how to employ that new model in controlling Khepera robot and mapping the vision information into motor speeds to develop obstacle avoidance navigation that depends only on the vision information.

4.6.1 Experiment 8: Evolving Vision Based Navigation

This experiment was performed by Floreano and Mattiussi [22]. In the experiment, the robot was placed in a rectangular arena whose walls are covered with vertical white and black strips with variable width. The Khepera robot is provided with K213 vision module similar to the one described in the co-evolution experiment in section 3.5. The goal of the evolved controller is to use the information available from the vision module to enable the robot to navigate without colliding with the walls.

The vision module provides a linear image consisting of 64 pixels that cover an angle of 36 degrees. Only 16 equally spaced photoreceptors are used as inputs to the spiking neural network. The values of photoreceptors readings are filtered to obtain information about the contrast, scaled to the range of

[0, 1] and then sent to the spiking neural network. There are extra 2 input neuron in the network whose input is the difference between the actual and the desired motor speeds. Again this difference is scaled to the rage of [0, 1] before being sent to the spiking neural network. The network contained four output neurons, two for each motor speed. The two neurons set the forward and backward speed for each motor. The actual speed sent to the motor is their algebraic sum. In addition to the 18 input neurons and the 4 output neurons the network contained 10 neurons that are connected to the input and output neurons.

The input vision photoreceptors and the output motor speed are interfaced to the spiking neural network as follows. The 16 scaled inputs of photoreceptors are used to set the probability to emit a spike by the corresponding input neurons. Also, the firing rates of the 4 output neurons are mapped to the motor speeds. This explains the reason of using two neurons for each motor speed since that firing rate of the output neurons can not take negative values. The cycle of reading the photoreceptors and updating the motor speed goes in the following order. Every 100 ms, the input photoreceptors are read, filtered, scaled and used to set the probability of emitting a spike by the input neurons. During the 100 ms cycle, the activation level of each neuron, except input neurons, is updated every 1 ms according to the model of equation (4.5) and the neurons are allowed to emit spikes if their activation level exceeds the threshold. At the end of the 100 ms cycle, the spiking rate of the output neurons, calculated over the last 20 ms period of the cycle, is used to update the motor speeds.

The genetic algorithm is used to obtain the best synaptic weights connecting the spiking neurons. The population consisted of 60 individuals and the experiment lasted for 30 generations. Each individual is tested in 400 cycle, in which its fitness is the sum of its motor speeds if they are both positive and zero otherwise. This fitness function will reward the individuals that move forward while offering no reward to individuals that rotate (due to difference in the sign of the speeds) or move backward (when both speeds are negative). The fitness evaluation of the individual is the average of its fitness over the 400 cycles.

The results of the experiment showed that the best individual was able to move in curved trajectories of large radii but without colliding with the surrounding walls. The experiment was repeated using a classical neural network with sigmoid activation function and with same architecture. However, the fitness of its individuals didn't increase with time and its individual neural network controllers were not able to map the vision information into motor speed that secure a safe navigation without colliding with the surrounding walls [22].

4.7 Comment on different approaches of evolutionary robotics

We presented different approaches for evolving controllers such as neural networks, fuzzy logic and spiking neural networks. Each approach has appealing advantages as one form of controller for mobile and autonomous robots. It may also include some difficulties or limitations when being evolved. We try in this section to shed some light on the attractive features of these different approaches and some issues that need to be considered when combined with evolutionary computations.

As a general approach, fuzzy logic provides a tool for dealing with systems with uncertain models which suits the dynamic and possibly unknown environments encountered by mobile robots. It has the advantages of implementing human knowledge. It simulates the human method of reasoning by using linguistic variables and knowledge that is represented by its rule base. For example, the human experience in walking or navigation while avoiding possible obstacles can be moved to the robot brain through using a fuzzy controller whose rules are based on this experience.

Another useful feature of fuzzy logic that is interesting in the field of robotics is its ability to combine different rules outputs in the defuzzification process. This ability can be further used in behavior coordination. In this approach different controllers are designed independently, possibly by fuzzy logic, neural networks or even designed by human programmers. Every controller implements a certain behavior or task. A simple example is two controllers for obstacle avoidance and goal seeking. Our problem in behavior coordination is to combine results from different behaviors in one command to send to the effectors or motors. The fuzzy approach for this problem works by providing a number of rules that assigns weights for fusing the different outputs from the controllers based on the current situation. In our example, a typical rule will favor the output of obstacle avoidance behavior when a near obstacle is detected. This method provides a way of combining the outputs of many behaviors each control cycle unlike behavior arbitration methods that choose one active behavior each time based on fixed or dynamic priorities. As we mentioned, these rules can be based on human experience. Further more, genetic algorithm can be employed to evolve the best set of rules for behavior coordination. In fact, this was the approach used by Tunstel et al. in [23] to evolve fuzzy behavior arbitration for planetary microrovers.

On the other hand, fuzzy logic approach lacks a standard method for creating the rules based on the human experience. Also the time taken in computations especially in the defuzzification process may affect the real time performance of the controller and the the robot if not performed using dedicated processors [24]. Another issue that needs to be considered when designing fuzzy logic controller for a robot is the design of the membership functions. In some experiments, redesigning the membership functions led to avoiding oscillations in the robot behavior [25].

Evolutionary computation appears to be a good solution to the problem of automatic design of the fuzzy logic controller. However there are some issues that the controller designer should consider when evolving the fuzzy logic controller. One of these issues is deciding what to evolve, whether it is the membership function parameters, the rules or both of them. Evolving both rules and membership functions has the advantage of decreasing chances of errors due to miss choices made in the early stages of the design, however the evolution process will search in a larger space for the best set of rules and best parameters for the membership functions. It should be noted that even by evolving the rules and the membership parameters, this can not eliminate the designer choice of the type of membership function (triangular or trapezoidal ...etc). Evolving the fuzzy behavior coordination module mentioned earlier is an example of evolving the fuzzy rules while the experiment in section four of this chapter is an example of evolving the fuzzy rules along with the membership functions.

Another issue to be considered in evolving fuzzy logic controller is the number of rules. The number of rules can affect the speed and performance of the robot and the choice of the genetic algorithm as well. Small number of rules will decrease the computations in the fuzzy logic controllers but on the other hand this small number may not cover all the possible situations or sensors combinations encountered by the robot. Evolving controller with fixed number of rules or fixed maximum number of rules will lead to using fixed length chromosome. The other approach of using population of individuals with different number of rules requires variable length chromosomes and possible modification of the genetic operator. Messy genetic algorithm [27] can be a potential evolutionary algorithm for evolving the fuzzy logic controller with variable number of rules. It has a modified version of the traditional crossover genetic operator called cut and slice operator that can deal with the variation of the genetic material length. In fact, it was used by Hoffman and Pfister in [26] to evolve the rules for fuzzy logic controllers to enable a mobile robot to reach its target while avoiding the obstacles.

Another approach of evolutionary robotics that we presented is evolving neural networks. Artificial neural networks offer many characteristics that make them suitable for the problem of controlling autonomous robots. First, the noise present in the sensor readings, whether they are sonar sensors or infrared sensors, makes the neural networks suitable controllers due to their known tolerance to noise. Moreover, if one of the sensors was not functioning, the output of the neural network could still be acceptable [7]. Second, the neural networks are able to learn and they could be trained. The weights and the thresholds and other parameters of the neural network could be adjusted to produce different behaviors even for the same network architecture. Also, neural networks can select the sensors that are suitable for a given behavior by adjusting the weight corresponding to each sensor or input.

As in the case of fuzzy logic, the genetic algorithm can offer an automatic way for designing neural network controller by evolving the synaptic weights or the network architecture or both of them. Neural networks have many existing learning algorithms, but the genetic algorithms offers potential advantage of the parallel search by using a population of individuals. An issue to be considered in evolving neural networks that may affect the genetic algorithm is the size of the parameter to be evolved. Large networks with large number of synaptic weights may require a long chromosome. In this case the real encoding of these parameters could be considered instead of the binary encoding.

Compared to fuzzy logic, the learning of the neural network which is stored as synaptic weights can not be acquired by human reasoning [24]. For example, in the experiment of trash collection no direct relation was found between modules of the neural network and the certain behavior of the robot, sometimes by observing the activation level of some neurons and certain behaviors of the robot we could find a correlation as in the experiment of home seeking but this is not the general case. On the other hand, the knowledge represented by the rules of the fuzzy logic controller can be acquired by human reasoning. For example, we could read on of the evolved rules in the experiment of evolving fuzzy logic controller and understand what it implies. Another point is that we can not easily implement high level behavior using neural networks as we can do using fuzzy logic. Although many relatively complex behaviors have be evolved using neural networks, such as trash collection, implementing a high level reasoning and selection or coordination between behaviors would require a method that mimics human reasoning.

We have also presented in this chapter a relatively new approach in evolutionary robotics which is evolving behaviors using spiking neural networks. The dynamic model of the spiking neural network suits the time changing relation between the sensors and the motors [22]. On the other hand, the complexity of the model and the need of the interface between the sensors of the robot and input of the spiking neural network have limited the experiments of evolutionary robotics that use it compared to other widely used approaches as artificial neural networks or fuzzy logic. Analog Very Large Scale Integrated Circuits (VLSI) can implement spiking neural networks using circuits with very small area and power consumption, which is an advantage over other approaches. In [28], an analog VLSI circuit that implemented spiking neural networks was used for controlling a robotic leg.

To summarize, each approach of evolutionary robotics is characterized by some potential advantages that makes it a suitable solution for the problem of controlling mobile robots. Also each approach has some limitations or difficulties when being evolved. Choosing which approach is a trade off between the advantages and the limitations.

4.8 Summary

In the previous sections we have seen how the evolutionary computations algorithms were successfully used to evolve many types of controllers for Khepera robot. It was used to evolve neural network synaptic weights in the obstacle avoidance behavior of experiment 1 and the battery recharging behavior of experiment 3. We have also seen how it can evolve the architecture of the neural network along with the synaptic weights as in the experiment of evolving light seeking behavior. Alternatively, it can evolve the learning rules and learning rate necessary for training the neural network synaptic weights. Other types of controllers were successfully evolved too, such as fuzzy logic controllers and computer programs.

Many other experiments are conducted using evolutionary computations on different robotic platforms recently. In fact, evolutionary computation is a very promising approach for designing controllers for mobile robots.

Acknowledgement

The author would like to thank S. Mercorious and S. Kirolos for their support all over the past years.

References

1. K-Team, "Khepera User Manual," Lasuanne, Switzerland, 1999.
2. F. Mondada, F. Franz and I. Paolo, "Mobile Robot Miniaturisation: A Tool for Investigation in Control Algorithm," Proceedings of the Third International Symposium on Experimental Robotics, Kyoto, Japan, 1993.
3. A. Schultz and J. Grefenstette, "Using a Genetic Algorithm to Learn Behaviors for Autonomous Vehicles," Naval Research Laboratory, Washington, Dc, 1992.
4. J. Meyer, P. Husbands and I. Harvey, "Evolutionary Robotics: a Survey of Applications and Problems," In Evolutionary Robotics : First European Workshop, Evorobot'98, P. Husbands and J. Meyer (editors), Springer Verlag 1998.
5. I. Harvey, P. Husbands, D. Cliff, A. Thompson, N. Jakobi, "Evolutionary Robotics: the Sussex Approach," In Robotics and Autonomous Systems, Vol. 20, pp. 205-224, 1997.
6. A. Loffler, J. Klahold and U. Ruckert, "The Mini-Robot Khepera as a Foraging Animate: Synthesis and Analysis of Behavior," In Proceedings of the Fifth International Heinz Nixdorf Symposium: Autonomous Minirobots for Research and Edutainment (AMiRE), Vol. 97, pp. 93-130, 2001.
7. D. Floreano and F. Mondada, "Automatic Creation of An Autonomous Agent: Genetic Evolution of a Neural Network Driven Robot," From Animals to Animats:3, Proceedings of the Conference on Simulation of Adaptive Behavior, edited by D. Cliff, P. Husbands and S. Wilson, MIT Press, 1994.
8. M. Botros "Evovlving Neural Network Based Controllers for Autonomous Robots Using Genetic Algorithms," Master Thesis, Cairo University, Egypt, 2003.

9. M. Hulse, B. Lara, F. Pasemann and U. Steinmetz, "Evolving Neural Behaviour Control for Autonomous Robots," Max-Planck Institute for Mathematics in the Sciences, Leipzig, Germany, 2001.
10. D. Floreano and F. Mondada, "Evolution of Homing Navigation in a Real Mobile Robot," IEEE Transactions on Systems, Man, and Cybernetics (B), Vol. 2, pp. 396-407, 1996.
11. S. Nolfi, "Using Emergent Modularity to Develop Control Systems for Mobile Robots," Journal of Adaptive Behavior, Vol. 5, pp. 343-363, 1997.
12. C. Scheier and R. Pfeifer, "Classification as Sensory-Motor Coordination," Advances in Artificial Life: Proceedings of the Third European Conference on Artificial Life, edited by F. Moran, A.Moreno, J. Merelo and P. Chacon, Springer Verlag, 1995.
13. D. Floreano and S. Nolfi, "God Save the Red Queen! Competetion in Coevolutionary Robotics," Genetic Programming 1997: Proceedings of the Second Annual Conference, Stanford University, edited by J. Koza, K. Deb, M. Dorigo, D. Fogel, M. Garzon, H. Iba, and R. Riolo, pp. 398-406, 1997.
14. D. Floreano and S. Nolfi, "Adaptive Behavior in Competing Co-Evolving Species," Fourth European Conference on Artificial Life, MIT press, Cambridge MA, editted by P. Husbands and I. Harvey, pp. 378-387, 1997.
15. Zadeh, L., "Outline of a New Approach to the Analysis of Complex Systems and Decision Process," IEEE Transaction Systems, Man and Cybernetics, Vol. 3, pp 28-40, 1973.
16. E. Mamdani and S. Assilian, "An Experiment in Linguistic Synthesis with Fuzzy Logic Controller," Journal of Man-Machine Studies, Vol. 7, pp. 1-7, 1975.
17. S. Lee and S. Cho, "Emergent Behaviors of a Fuzzy Sensory-Motor Controller Evolved by Genetic Algorithm," IEEE Transaction Systems Man and Cybernetics (B), Vol. 31, No. 6, pp. 919-929, 2001.
18. J. Koza, "Genetic Programming," MIT Press, Cambridge MA, 1992.
19. P. Nordin and W. Banzhaf, "Genetic Programming Controlling a Miniature Robot," Working Notes for the AAAI Symposium on Genetic Programming, MIT, Cambridge MA, 1995.
20. W. Gerstner and W. Kistler, "Spiking Neuron Models," Cambridge University Press, 2002.
21. W. Gerstner, J. van Hemmen, and J. Cowan, "What Matters in Neuronal Locking?," Neural Computation, Vol. 8, pp. 1653-1676, 1996.
22. D. Floreano and C. Mattiussi, "Evolution of Spiking Neural Controllers for Autonomous Vision-Based Robots," Evolutionary Robotics. From Intelligent Robotics to Artificial Life, Springer Verlag, Tokyo, 2001.
23. E. Tunstel, H. Danny, and M. Jamshidi, " Behavior Hierarchy for Autonomous Mobile Robots: Fuzzy-behavior modulation and evolution," International Journal of Intelligent Automation and Soft Computing, Special Issue: Autonomous Control Engineering at NASA ACE Center, Vol. 3, pp. 37-49, 1997.
24. J. Godjevac, "Comparative Study of Fuzzy Control, Neural Network Control and Neuro-Fuzzy Control", In Fuzzy Set Theory and Advanced Mathematical Applications,D. Ruan Ed., Kluwer Academic, Chapter 12, pp. 291-322, 1995.
25. S. Marapane, M. Trivedi, N. Lassiter and M. Holder, "Motion Control of Cooperative Robotic Teams through Visual Observation and Fuzzy Logic Control," Proceedings of IEEE International Conference on Robotics and Automation, Vol. 2, pp. 1738-1743, 1996.

26. F. Hoffmann and G. Pfister, “Evolutionary Design of a Fuzzy Knowledge Base for a Mobile Robot,” International Journal of Approximate Reasoning, Vol. 17, pp. 447-469, 1997.
27. D. Goldberg, B. Krob and K. Deb, “Messy Gentic Algorithms Motivations, Analysis and First Results,” Complex Systems, Vol. 3, pp. 493-530, 1989.
28. M. A. Lewis, M. Hartmann, R. Etienne-Cummings, and A. Cohen, “Biomorphic Control of a Running Robot Leg using a Custom aVLSI CPG Chip,” Neurocomputing, Vol. 38-40, pp. 1409-1421, June 2001.

Part II

Evolvable Hardware Synthesis

5

Evolutionary Synthesis of Synchronous Finite State Machines

Nadia Nedjah and Luiza de Macedo Mourelle

Department of System Engineering and Computation,
Engineering Faculty,
State University of Rio de Janeiro,
Rua São Francisco Xavier, 524, Sala 5022-D,
Maracanã, Rio de Janeiro, Brazil
(nadia | ldmm)@eng.uerj.br
www.eng.uerj.br

Synchronous finite state machines are very important for digital sequential designs. Among other important aspects, they represent a powerful way for synchronising hardware components so that these components may cooperate adequately in the fulfilment of the main objective of the hardware design. In this chapter, we propose an evolutionary methodology synthesise finite state machines. First, we optimally solve the state assignment NP-complete problem, which is inherent to designing any synchronous finite state machines using genetic algorithms. This is motivated by the fact that with an optimal state assignment one can physically implement the state machine in question using a minimal hardware area and response time. Second, with the optimal state assignment provided, we propose to use the evolutionary methodology to yield optimal evolvable hardware that implement the state machine control component. The evolved hardware requires a minimal hardware area and introduces a minimal propagation delay of the machine output signals.

5.1 Introduction

Sequential digital systems or simply finite state machines have two main characteristics: there is at least one feedback path from the system output signal to the system input signals; and there is a memory capability that allows the system to determine current and future output signal values based on the previous input and output signal values [15].

Traditionally, the design process of a state machine passes through five main steps, wherein the second and third steps may be bypassed as shown in Fig. 5.1:

1. the specification of the sequential system, which should determine the next states and outputs of every present state of the machine. This is done using state tables and state diagrams;
2. the state reduction, which should reduce the number of present states using equivalence and output class grouping;
3. the state assignment, which should assign a distinct combination to every present state. This may be done using Armstrong-Humphrey heuristics [15];
4. the minimisation of the control combinational logic using K-maps and transition maps;
5. finally, the implementation of the state machine, using gates and flip-flops.

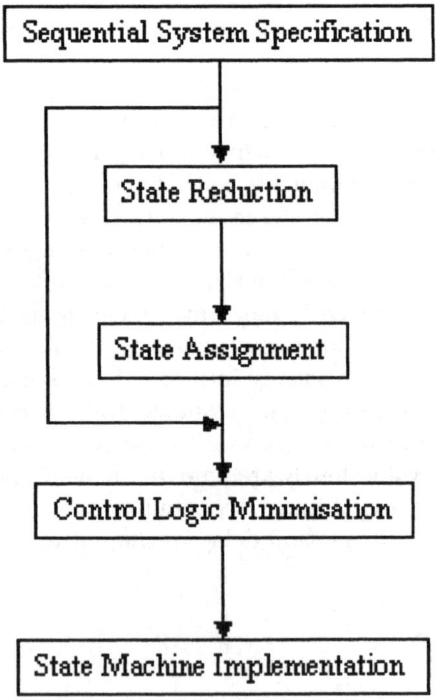

Fig. 5.1. The structural description of a finite synchronous state machine

In this chapter, we concentrate on the third and forth steps of the design process, i.e. the state assignment problem and the control logic minimisation. We present a genetic algorithm designed for finding a state assignment of a given synchronous finite state machine, which attempts to minimise the cost related to the state transitions. Then, we use genetic programming to evolve the circuit that controls the machine current and next states.

The remainder of this chapter is organised into seven sections. In Section 5.2, we introduce the problems that face the designer of finite state machine, which are mainly the state assignment problem and the control logic. We show that a better assignment improves considerably the cost of the control logic. In Section 5.3, we give a thourough overview on the principles of evolutionary computations and genetic algorithms and their application to solve NP-problems. In Section 5.4, we design a genetic algorithm for evolving best state assignment for a given state machine specification. We describe the genetic operators used as well as the fitness function, which determines whether a state assignment is better that another and how much. In Section 5.5, we present results evolved through our genetic algorithm for some well-known benchmarks. Then we compare the obtained results with those obtained by another genetic algorithm described in [1] as well as with NOVA, which is uses well established but non-evolutionary method [16]. In Section 5.6, we briefly introduce the genetic programming concepts and their applications to engineer evolvable hardware. Subsequently, we present a genetic programming-based synthesiser for evolving minimal control logic circuit provided the state assignment for the specification of the state machine in question. We describe the circuit encoding, genetic operators used as well as the fitness function, which determines whether a control logic design is better than another and how much. In Section 5.7, we compare the are and time requirements of the designs evolved through our evolutionary synthesiser for some well-known benchmarks and compare the obtained results with those obtained using the traditional method to design state machine, i.e. using Karnaugh maps and flip-flop transition maps. In Section 5.8, we summarise the ideas presented throughout the chapter and draw some conclusions.

5.2 Synchronous Finite State Machines

Once the specification and the state reduction step have been completed, the next step consists then of assigning a code to each state present in the machine. It is clear that if the machine has N distinct states then one needs N distinct combinations of 0s and 1s. So one needs K flip-flops to store the machine current state, wherein K is the smallest positive integer such that $2^K \geq N$. The state assignment problem consists of finding the best assignment of the flip-flop combinations to the machine states. Since a machine state is nothing but a counting device, combinational control logic is necessary to activate the flip-flops in the desired sequence. This is shown in Fig. 5.2, wherein the feedback signals constitute the machine state, the control logic is a combinational circuit that computes the state machine output signals (also called *primary output signals*) from the state signals (also called *current state*) and the input signals (also called *primary input signals*). It also produces the signals of new machine state (also called *next state*).

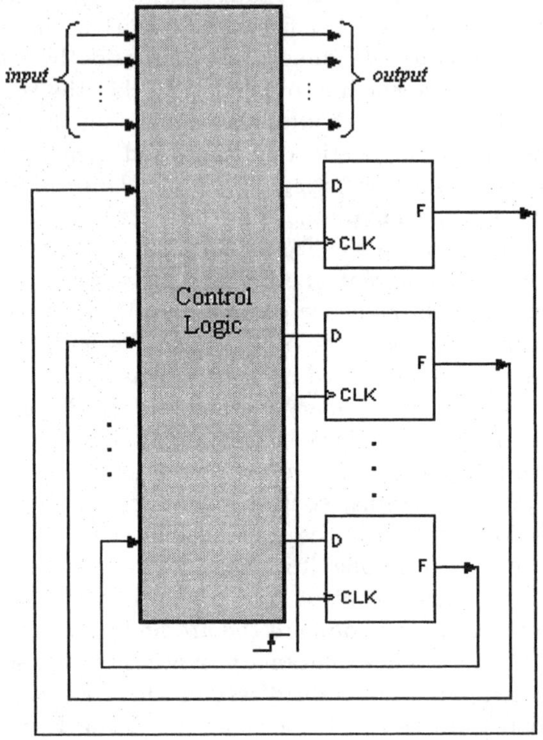

Fig. 5.2. The structural description of a finite synchronous state machine

The control logic component in a state machine is responsible of generating the primary output signals as well as the signal that form the next state. It does so using the primary input signals and the signals that constitute the current state (see Fig. 5.2). Traditionally, the combinational circuit of the control logic is obtained using the transition maps of the flip-flops [15]. Given a state transition function, it is expected that the complexity (area and time) and so the cost of the control logic will vary for different assignments of flip-flop combinations to allowed states. Consequently, the designer should seek the assignment that minimises the complexity and so the cost of the combinational logic required to control the state transitions.

5.2.1 Example of State Machine

Consider the state machine of one input signal (I), one output signal (O) and four states whose state transition function is given in tabular form in Table

5.1 and assume that we use D-flip-flops to store the machine current state. Then the state assignment $A_0 = \{s_0 \equiv 00, s_1 \equiv 11, s_2 \equiv 01, s_3 \equiv 10\}$ requires a control logic that consists of three AND gates, five AND gates and three OR gates while the assignments $A_1 = \{s_0 \equiv 00, s_1 \equiv 10, s_2 \equiv 01, s_3 \equiv 11\}$ requires a control logic that consists of only two NOT gates, five AND gates and two OR gates. The schematics of the state machines that encode the state according to state assignments A_0 and A_1 are given in Fig. 5.3 and Fig. 5.4 respectively.

Table 5.1. Example of state transition function

Present State	Next State		Output (O)	
	$I = 0$	$I = 0$	$I = 1$	$I = 1$
q_0	q_0	q_0	0	0
q_1	q_2	q_2	0	1
q_2	q_0	q_0	1	0
q_3	q_2	q_2	1	1

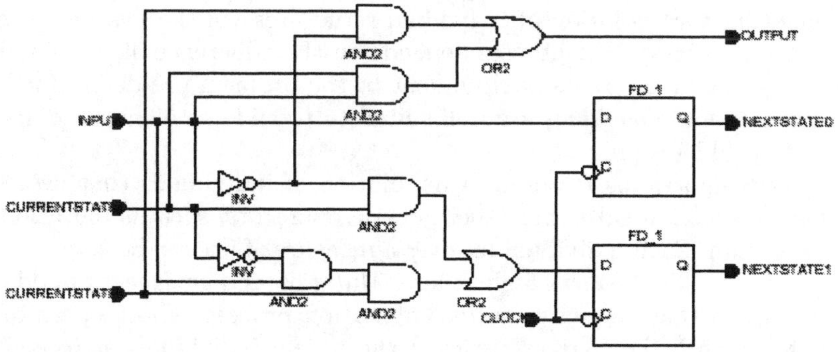

Fig. 5.3. The machine state schematics for state assignment A_0

In Section 5.3, we concentrate on the third step of the design process, i.e. the state assignment problem. We present a genetic algorithm designed for finding a state assignment of a given synchronous finite state machine, which attempts to minimise the cost related to the state transitions. In Section 5.5, we focus on evolving minimal control logic for state machines, provided the state assignment.

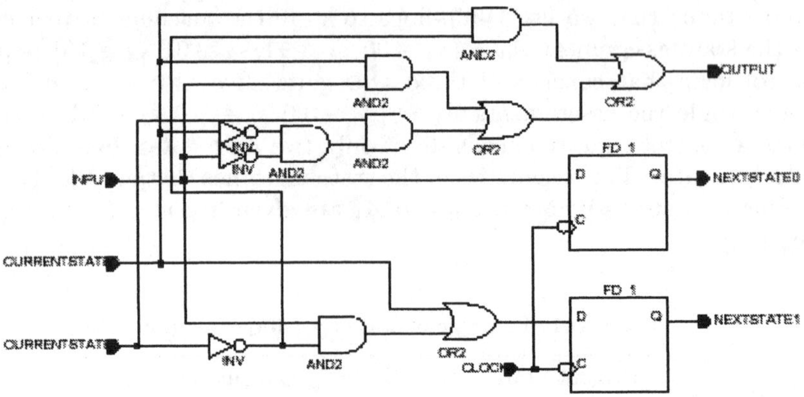

Fig. 5.4. The machine state schematics for state assignment A_1

5.3 Principles of Genetic Algorithms

Evolutionary algorithms are computer-based solving systems, which use the evolutionary computational models as key element in their design and implementation. A variety of evolutionary algorithms have been proposed. The most popular ones are genetic algorithms [13]. They have a conceptual base of simulating the evolution of individual structures via the Darwinian natural selection process. The process depends on the adherence of the individual structures as defined by its environment to the problem pre-determined constraints. *Genetic algorithms* are well suited to provide an efficient solution of NP-hard problems [4].

Genetic algorithms maintain a population of individuals that evolve according to selection rules and other genetic operators, such as mutation and recombination. Each individual receives a measure of fitness. Selection focuses on individuals, which shows high fitness. *Mutation* and *crossover* provide general heuristics that simulate the recombination process. Those operators attempt to perturb the characteristics of the parent individuals as to generate distinct offspring individuals.

Genetic algorithms are implemented through the following generic algorithm described by Algorithm 5.1, wherein parameters ps, f and gn are the *population size*, *fitness* of the expected individual and the number of *generation* allowed respectively.

In Algorithm 5.1, function *intialPopulation* returns a valid random set of individuals that compose the population of the first generation, function *evaluate* returns the fitness of a given population. Function *select* chooses according to some criterion that privileges *fitter* individuals, the individuals that will be used to generate the population of the next generation and function reproduction implements the crossover and mutation process to yield the

Algorithm 5.1 Genetic Algorithms

input: population size (ps), expected fitness (f), last generation number (gn);
output: fittest individual (fit);

```
generation := 0;
population := initialPopulation();
fitness := evaluate(population);
do
    parents := select(population);
    population := reproduce(parents);
    fitness := evaluate(population);
    generation := generation + 1;
    fit := fittestIndividual(population);
while(fit < f) and (generation < gn);
End.
```

new population. The main genetic operators will be described in the following sections.

5.3.1 Assignment Encoding

Encoding of individuals is one of the implementation decisions one has to make in order to use genetic algorithms. It very depends on the nature of the problem to be solved. There are several representations that have been used with success [13]: *binary encoding* which is the most common mainly because it was used in the first works on genetic algorithms, represents an individual as a string of bits; *permutation encoding* mainly used in ordering problem, encodes an individual as a sequence of integer; *value encoding* represents an individual as a sequence of values that are some evaluation of some aspect of the problem; *tree encoding* represents an individual as a tree. This encoding is generally used to represent structured individuals such as computer programs, mathmatical expressions and circuits.

5.3.2 Individual Reproduction

Besides the parameters which represent the population size, the fitness of the expected result and the maximal number of generation allowed, the genetic algorithm has several other parameters, which can be adjust by the user so that the result is up to his or her expectation. The *selection* is performed using some selection probabilities and the *recombination*, as it is subdivided into *crossover* and *mutation* processes, depends on the kind of crossover and the mutation *rate* and *degree* to be used.

Selection

The selection problem consists of how to select the individuals that should yield the new population. According to Darwins evolution theory the best ones should survive longer and create more new offspring. There are many selection methods [6], [9]. These methods include *roulette wheel* selection or *fitness proportionate* reproduction and *rank* selection. In the following, we describe the idea behind each of these selection methods. In our implementation, we use fitness proportionate reproduction.

In fitness proportionate reproduction, parents are selected according to their fitness. The better the fitness the individuals have, the higher their chances to be selected are. Imagine a roulette wheel where are placed all individuals of the population, wherein every individual has portion proportionate to its fitness, as it is shown in Fig. 5.5.

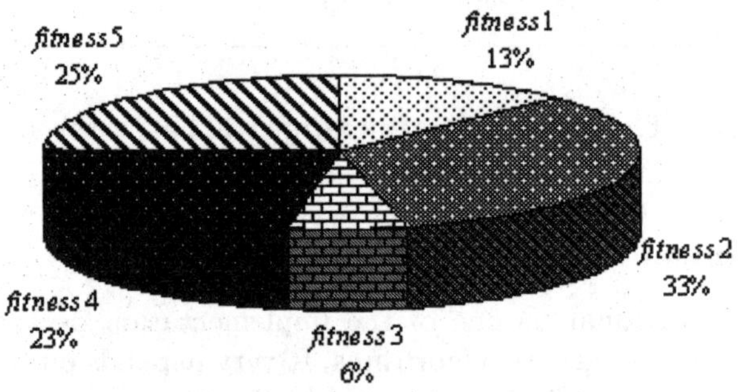

Fig. 5.5. Representation with the roulette wheel selection

Then a marble is thrown into the roulette and selects an individual. It is clear that individuals with bigger portion in the wheel will be selected more times. The selection process can be simulated by following steps:

1. first, sum up the fitness of all individuals in the population and let S be the obtained sum;
2. then generate a random number from the $[0, S]$, and let f be this number;
3. subsequently, go through the individuals of the population, summing up the fitness of the next one. Let σ be this partial sum;
4. if $\sigma \geq f$, then stop the selection process and choose the current individual otherwise return to second step.

The fitness proportionate reproduction selection presents some limitations when the individual fitnesses differ too much from one another. For instance, if the best individual has a fitness of 95% of the entire roulette wheel then the other individuals will have very few, if any, chances to be selected. To get round this limitation, the rank selection method first ranks the individuals of the population according to their corresponding fitnesses. The individual with the worst fitness receives $rank_1$ and that with the best fitness receives $rank_N$, which is the number of individuals in the population. The impact of the ranking process is shown in Fig. 5.6, which represents the roulette wheel before and after the ranking process. Rank selction may yield a slower convergence as the fittest individuals and those that are less fit have much closer ranks.

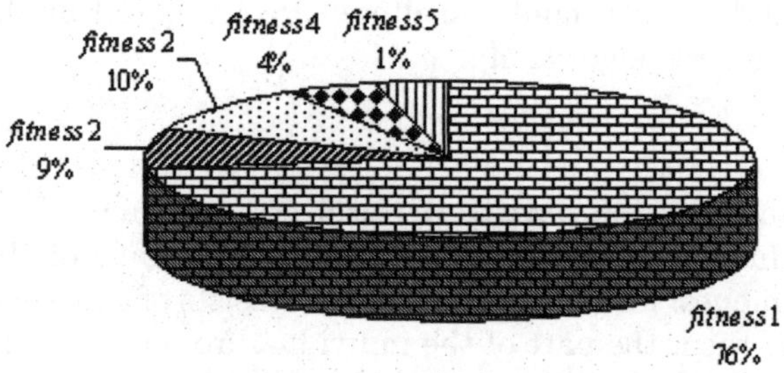

(a) before individual ranking

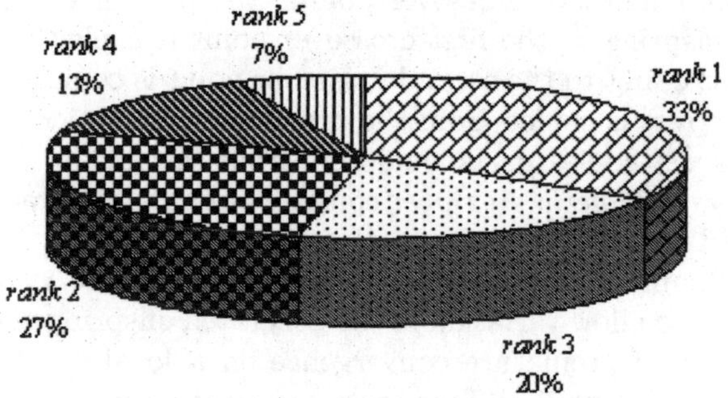

(b) after individual ranking

Fig. 5.6. Representation of the roulette wheel selection before and after ranking the individuals according to their fitnesses

Reproduction

Given the parents populations, the reproduction can proceed using different schemes [6], [9]: a *total replacement*, *steady-state replacement* and *elitism*. In the first scheme, offspring replace their parents in the population of the next generation. That is only offspring are used to form the population of the next generation. The steady-state replacement exploits the idea that only few low-fitness individuals should be discarded in the next generation and should then be replaced by offspring. Finally, elitism exploits the idea that the best solution might be the fittest individual of the current population and so transports it unchanged into the population of the next generation. In our implementation we use the total replacement reproduction scheme as well as elitism.

Obtaining offspring that share some traits with their corresponding parents is performed by the crossover function. There are several types of crossover operators. These will be presented shortly. The newly obtained population can then suffer some mutation, i.e. some of the individuals of some of the genes. The crossover type, the number of individuals that should be mutated and how far these individuals should be altered are set up during the initialisation process of the genetic algorithm.

Crossover

There are many ways on how to perform crossover and these may depend on the individual encoding used [13]. We present some of these techniques crossover techniques. *Single-point* crossover consists of choosing randomly one *crossover point* then, the part of the individual from the beginning of the offspring till the crossover point is copied from one parent, the rest is copied from the second parent as depicted in Fig. 5.7(a). *Double-point crossover* consists of selecting randomly two crossover points, the part of the individual from beginning of offspring to the first crossover point is copied from one parent, the part from the first to the second crossover point is copied from the second parent and the rest is copied from the first parent as depicted in Fig. 5.7(b). *Uniform crossover* copies parts randomly from the first or from the second parent. Finally, *arithmetic crossover* consists of applying some arithmetic operation to yield a new offspring.

The single-point and double-point crossover may use randomly selected crossover points to allow variation in the generated offspring and to contribute in the avoidance of premature convergence on a local optimum [5]. In our implementation, we tested all four-crossover strategies.

Mutation

Mutation consists of altering some genes of some individuals of the population obtained after crossover. The number of individuals that should be mutated is given by the parameter *mutation rate* while the parameter *mutation degree* states how many genes of a selected individual should be changed. The mutation parameters have to be chosen carefully as if mutation occurs very often then the genetic algorithm would in fact change to random search [5]. When either of the mutation rate or mutation degree is null, the population is then kept unchanged, i.e. the population obtained from the crossover procedure represents actually the next generation population.

The essence of the mutation process depends on the encoding type used. When binary encoding is used, the mutation is nothing but a bit inversion of those bit genes that were randomised. When permutation encoding is used, the mutation is reduced to a permutation of some randomly selected integer genes. When value encoding is used, a very small value is added or subtracted from the randomised genes. When tree encoding is used, a content of a tree node is altered.

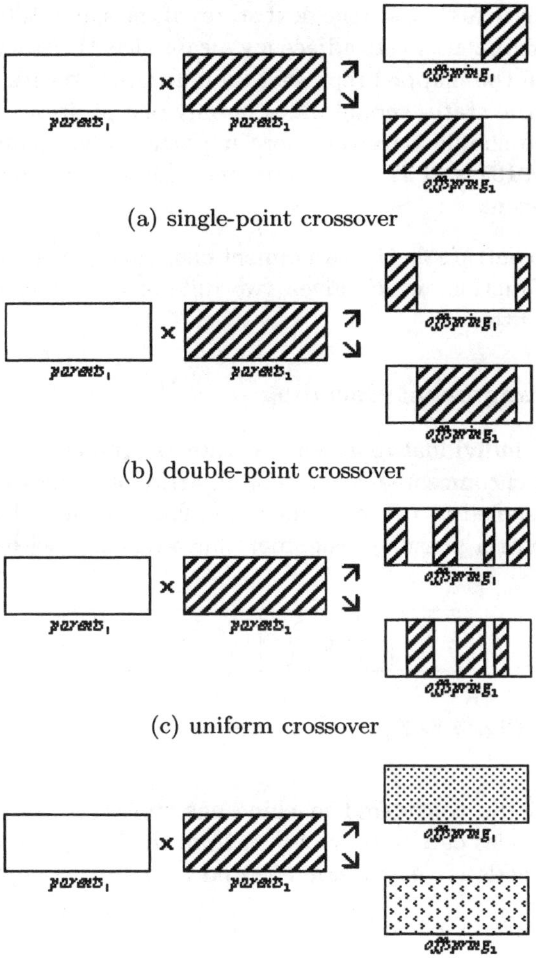

(a) single-point crossover

(b) double-point crossover

(c) uniform crossover

(d) arthmetic crossover

Fig. 5.7. Different types of crossover

5.4 Application to the State Assignment Problem

The identification of a good state assignment has been thoroughly studied over the years. In particular, Armstrong [2] and Humphrey [11] have pointed out that an assignment is good if it respects two rules, which consist of the following:

- two or more states that have the same next state should be given adjacent assignments;

- two or more states that are the next states of the same state should be given adjacent assignment. State adjacency means that the states appear next to each other in the mapped representation. In other terms, the combination assigned to the states should differ in only one position;
- the first rule should be given more important the second. For instance, state codes 0101 and 1101 are adjacent while state codes 1100 and 1111 are not adjacent.

Now we concentrate on the assignment encoding, genetic operators as well as the fitness function, which given two different assignment allows one to decide which is fitter.

5.4.1 State Assignment Encoding

In this case, an individual represents a state assignment. We use the integer encoding. Each chromosome consists of an array of N entries, wherein entry i is the code assigned to $ith.$ machine state. For instance, the chromosome in Fig. 5.5 represents a possible assignment for a machine with 6 states.

S_0	S_1	S_2	S_3	S_4	S_5
4	2	1	0	7	6

Fig. 5.8. Example of state assignment encoding

Note that if the considered machine has stores its state in K flip-flops, then the state codes can be only chosen from the integer interval $[0, 2^K - 1]$. Otherwise, the code is not considered valid as it can be kept in the machine memory.

5.4.2 Genetic Operators for State Assignments

As state assignments are represented using integer encoding, we could use single-point, double-point and uniform crossovers (see Section 5.3 for details). The mutation is implemented by altering a state code by another valid state. Note that when mutation occurs, a code might be used to represent two or more distinct states. Such a state assignment is not possible. In order to discourage the selection of such assignment, we apply a penalty every time a code is used more than once within the considered assignment. This will be further discussed in next section.

5.4.3 State Assinment Fitness Evaluation

This step of the genetic algorithm allows us to classify the individuals of a population so that fitter individuals are selected more often to contribute in the constitution of a new population. The fitness evaluation of state assignments is performed with respect to two rules of Armstrong [2] and Humphrey [11]:

- how much a given state assignment adheres to the first rule, i.e. how many states in the assignment, which have the same next state, have no adjacent state codes;
- how much a given state in the assignment adheres to the second rule, i.e. how many states in the assignment, which are the next states of the same state, have no adjacent state codes.

In order to efficiently compute the fitness of a given state assignment, we use an $N \times N$ *adjacency matrix*, wherein N is the number of the machine states. The triangular bottom part of the matrix holds the expected adjacency of the states with respect o the first rule while the triangular top part of it holds the expected adjacency of the states with respect to the second rule. The matrix entries are calculated as in Equation 5.1, wherein AM stands for the adjacency matrix, functions $next(\sigma)$ and $prev(\sigma)$ yield the set of states that are next and previous to state σ respectively. For instance, for the state machine in Table 5.2, we get the 4×4 adjacency matrix in Fig. 5.9.

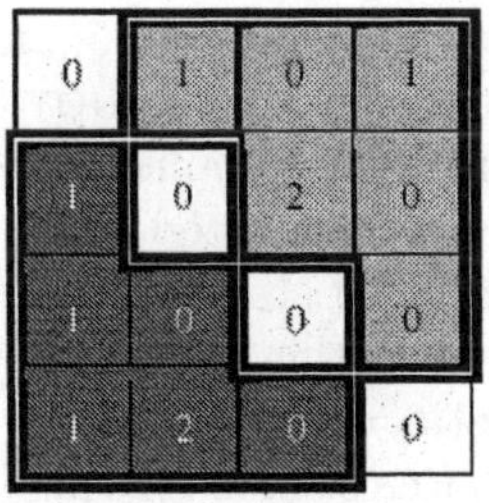

Fig. 5.9. Adjacency matrix for the machine state specified in Table 5.1

$$AM_{i,j} = \begin{cases} \#\left(next(q_i) \cup (next(q_j)\right) & \text{if } i > j \\ \#\left(prev(q_i) \cup (prev(q_j)\right) & \text{if } i < j \\ 0 & \text{if } i = j \end{cases} \tag{5.1}$$

Using the adjacency matrix AM, the fitness function applies a penalty of 2, respectively 1, every time the first rule, respectively the second rule, is broken. Equation 5.2 states the details of the fitness function applied to a state assignment σ, wherein function $na(q,p)$ returns 0 if the codes representing states q and p are adjacent and 1 otherwise. Note that state assignments that encode two distincts states using the same codes are penalised. Note that ψ represents the penalty.

$$fitness(\sigma) = \sum_{i \neq j \& \sigma_i = \sigma_j} \psi + \sum_{i=0}^{N-1} \sum_{j=i+1}^{N-1} (AM_{i,j} + 2 \times AM_{i,j}) \times na(\sigma_i, \sigma_j) \quad (5.2)$$

For instance, considering the state machine whose state transition function is described in Table 5.1, the state assignment $\{s_0 \equiv 00,\ s_1 \equiv 10,\ s_2 \equiv 01,\ s_3 \equiv 11\}$ has a fitness of 5 as the codes of states s_0 and s_3 are not adjacent but $AM - 0, 3 = 1$ and $AM_{3,0} = 1$ and the codes of states s_1 and s_2 are not adjacent but $AM_{1,2} = 2$ while the assignments $\{s_0 \equiv 00,\ s_1 \equiv 11,\ s_2 \equiv 01,\ s_3 \equiv 10\}$ has a fitness of 3 as the codes of states s_0 and s_1 are not adjacent but $AM_{0,1} = 1$ and $AM_{1,0} = 1$.

The objective of the genetic algorithm is to find the assignment that minimise the fitness function as described in Equation 5.2. Assignments with fitness 0 satisfy all the adjacency constraints. Such an assignment does not always exist.

5.5 Comparative Results

In this section, we compare the assignment evolved by our genetic algorithm to those yield by another genetic algorithm [5] and to those obtained using the non-evolutionary assignment system called NOVA [16]. The examples are well-known benchmarks for testing synchronous finite state machines [3]. Table 5.2 shows the best state assignment generated by the compared systems. The size column shows the total number of states/transitions of the machine.

Table 5.3 gives the fitness of the best state assignment produced by our genetic algorithm, the genetic algorithm from [1] and the two versions of NOVA system [16]. The *#AdjRes* stands for the number of expected adjacency restrictions. Each adjacency according to rule 1 is counted twice and that with respect to rule 2 is counted just once. For instance, in the case of the *Shiftreg* state machine, all 24 expected restrictions were fulfilled in the state assignment yielded by the compared systems. However, the state assignment obtained the first version of the NOVA system does not fulfil 8 of the expected adjacency restrictions of the state machine.

The chart of Fig. 5.10 compares graphically the degree of fulfilment of the adjacency restrictions expected in the state machines used as benchmarks. The chart shows clearly that our genetic algorithm always evolves a better state assignment.

5.6 Evolvable Hardware for the Control Logic

Genetic programming [10], [12] is way of producing a program using genetic evolution. The individuals within the evolutionary process are programs.

Table 5.2. Best state assignment yield by the compared systems for the benchmarks

FSM	System	State Assignment
Shiftreg	GA [1]	[0,2,5,7,4,6,1,3]
8/16	NOVA1	[0,4,2,6,3,7,1,5]
	NOVA2	[0,2,4,6,1,3,5,7]
	Our GA	[5,7,4,6,1,3,0,2]
Lion9	GA [1]	[0,4,12,13,15,1,3,7,5]
9/25	NOVA1	[2,0,4,6,7,5,3,1,11]
	NOVA2	[0,4,12,14,6,11,15,13,7]
	Our GA	[10,8,12,9,13,15,7,3,11]
Train11	GA [1]	[0,8,2,9,13,12,4,7,5,3,1]
11/25	NOVA1	[0,8,2,9,1,10,4,6,5,3,7]
	NOVA2	[0,13,11,5,4,7,6,10,14,15,12]
	Our GA	[2,6,1,4,0,14,10,9,8,11,3]
Bbarra	GA [1]	[0,6,2,14,4,5,13,7,3,1]
10/60	NOVA1	[4,0,2,3,1,13,12,7,6,5]
	NOVA2	[9,0,2,13,3,8,15,5,4,1]
	Our GA	[3,0,8,12,1,9,13,11,10,2]
Dk14	GA [1]	[0,4,2,1,5,7,3]
7/56	NOVA1	[5,7,1,4,3,2,0]
	NOVA2	[7,2,6,3,0,5,4]
	Our GA	[3,7,1,0,5,6,2]
Bbsse	GA [1]	[0,4,10,5,12,13,11,14,15,8,9,2,6,7,3,1]
16/56	NOVA1	[12,0,6,1,7,3,5,4,11,10,2,13,9,8,15,14]
	NOVA2	[2,3,6,15,1,13,7,8,12,4,9,0,5,10,11,14]
	Our GA	[15,14,9,12,1,4,3,7,6,10,2,11,13,0,5,8]
Donfile	GA [1]	[0,12,9,1,6,7,2,14,11,17,20,23,8,15,10,16,21,19,4,5,22,18,13,3]
24/96	NOVA1	[12,14,13,5,23,7,15,31,10,8,29,25,28,6,3,2,4,0,30,21,9,17,12,1]
	NOVA2	[6,30,11,28,25,19,0,26,1,2,14,10,31,24,27,15,12,8,29,23,13,9,7,3]
	Our GA	[2,18,17,1,29,21,6,22,7,0,4,20,19,3,23,16,9,8,13,5,12,28,25,24]

Table 5.3. Fitness of best assignments yield by the compared systems

State machine	*#AdjRes*	Our GA	GA [5]	NOVA1	NOVA2
Shiftreg	24	0	0	8	0
Lion9	69	21	27	25	30
Train11	57	18	19	23	28
Bbara	225	127	130	135	149
Dk14	137	68	75	72	76
Bbsse	305	203	215	220	220
Donfile	408	241	267	326	291

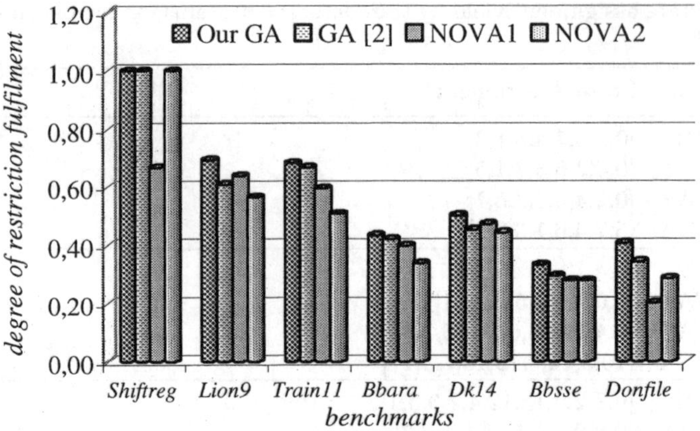

Fig. 5.10. Graphical comparison of the degree of fulfilment of rule 1 and 2 reached by the systems

The main goal of genetic programming is to provide a domain-independent problem-solving method that automatically yields computer programs from expected input/output behaviours. Exploiting genetic programming, we automatically generate novel control logic circuits that are *minimal* with respect to area and time requirements.

A circuit design may be specified using register-transfer level equations. Each instruction in the specification is an output signal assignment. A signal is assigned the result of an expression wherein the operators are those that represent basic gates in CMOS technology of VLSI circuit implementation and the operands are the input signals of the design. The allowed operators are shown in Table 5.4. Note that all gates introduce a minimal propagation delay as the number of input signal is minimal, which is 2.

Table 5.4. Gate name, symbol, gate-equivalent and propagation delay

Name	Symbol	Gate Code	Gate Equiv.	Delay
NOT		0	1	0.0625
AND		1	2	0.209
OR		2	2	0.216
XOR		3	3	0.212
NAND		4	1	0.13
NOR		5	1	0.156
XNOR		6	3	0.211
MUX		7	3	0.212

5.6.1 Circuit Encoding

We encode circuit designs using a matrix of cells that may be interconnected. A cell may or may not be involved in the circuit schematics. A cell consists of two inputs or three in the case of a MUX, a logical gate and a single output. A cell may draw its input signals from the output signals of gates of previous rows. The gates include in the first row draw their inputs from the circuit global input signal or their complements. The circuit global output signals are the output signals of the gates in the last raw of the matrix. An example of chromosome with respect to this encoding is given in Table 5.5. It represents the circuit of Fig. 5.11. Note that the input signals are numbered 0 to 3, their negated signals are numbered 4 to 7 and the output signals are numbered 16 to 19. If the circuit has n outputs with $n < 4$, then the signals numbered 16 to n are the actual output signals of the circuit.

Table 5.5. Chromosome for the circuit of Fig. 5.11

$\langle 1,0,2,8 \rangle$	$\langle 5,10,9,12 \rangle$	$\langle 7,13,14,11,16 \rangle$
$\langle 2,4,3,9 \rangle$	$\langle 1,8,10,13 \rangle$	$\langle 3,11,12,17 \rangle$
$\langle 3,1,6,10 \rangle$	$\langle 4,9,8,14 \rangle$	$\langle 7,15,14,15,18 \rangle$
$\langle 7,5,7,7,11 \rangle$	$\langle 4,10,11,15 \rangle$	$\langle 1,11,15,19 \rangle$

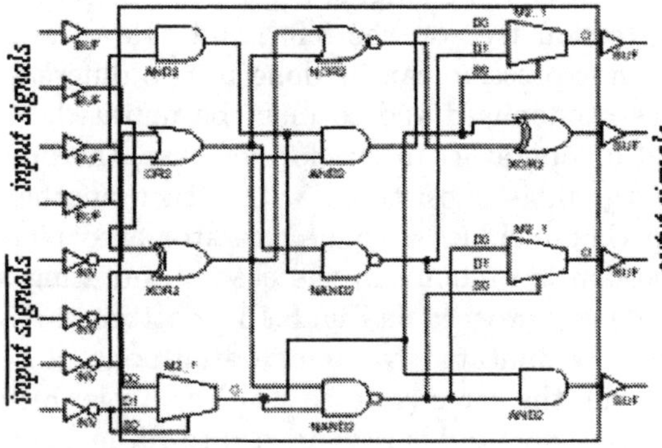

Fig. 5.11. Encoded circuit schematics

5.6.2 Circuit Reproduction

Crossover recombines two randomly selected circuits into two fresh offsprings. It may be single-point or double-point or uniform crossover as explained earlier. Crossover of circuit specification is implemented using a variable four-point crossover as described in Fig. 5.12.

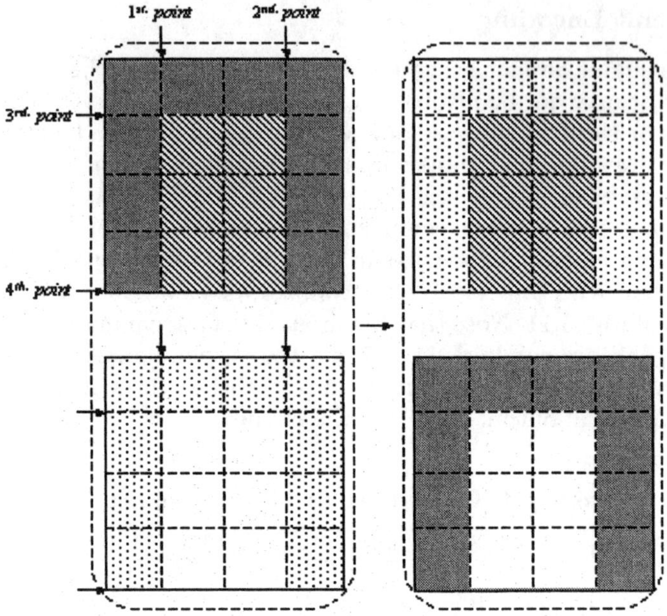

Fig. 5.12. Four-point crossover of circuit schematics

One of the important and complicated operators for genetic programming is the mutation. It consists of changing a gene of a selected individual. Here, a gene is the expression tree on the left hand side of a signal assignment symbol. Altering an expression can be done in two different ways depending the node that was randomised and so must be mutated. A node represents either an operand or operator. In the former case, the operand, which is a bit in the input signal, is substituted with either another input signal or simple expression that includes a single operator as depicted in Fig. 5.13 - top part. The decision is random. In the case of mutating an operand node to an operator node, we proceed as Fig. 5.13 - bottom part. The randomised operator node may be mutated to an operator node or to an operator of smaller (AND to NOT), the same (AND to XOR) or bigger arity (AND to MUX). In the last case, a new operand is randomised to fill in the new operand.

5.6.3 Circuit Evaluation

Another important aspect of genetic programming is to provide a way to evaluate the adherence of evolved computer programs to the imposed constraints. In our case, these constraints are of three kinds:

- First of all, the evolved specification must obey the input/output behaviour, which is given in a tabular form of expected results given the inputs. This is the truth table of the expected circuit.

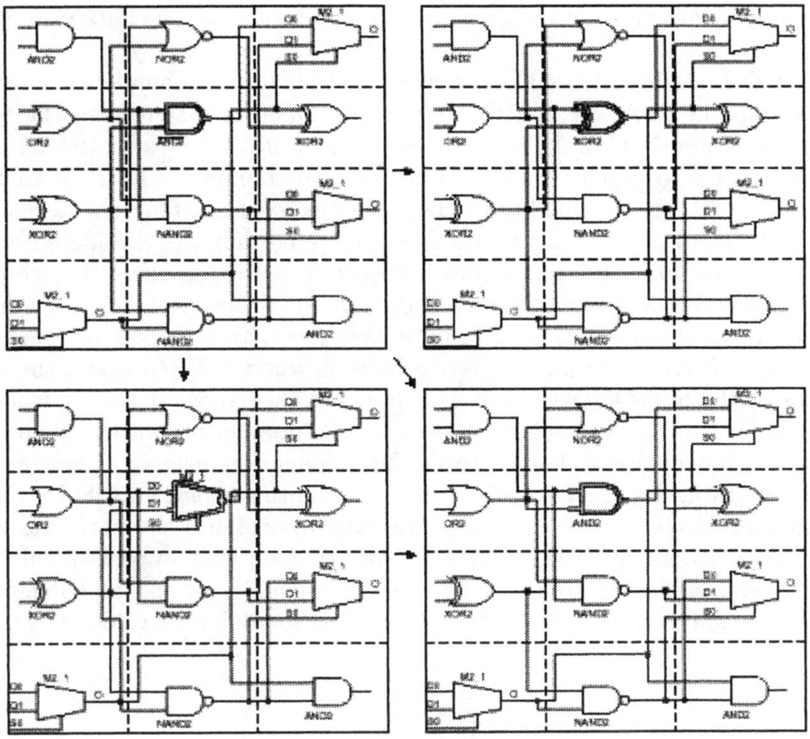

Fig. 5.13. Operand node mutation for circuit specification

- Second, the circuit must have a reduced size. This constraint allows us to yield compact digital circuits.
- Thirdly, the circuit must also reduce the signal propagation delay. This allows us to reduce the response time and so discover efficient circuits. In order to take into account both area and response time, we evaluate circuits using the weighted sum approach.

We estimate the necessary area for a given circuit using the concept of gate equivalent. This is the basic unit of measure for digital circuit complexity [7]. It is based upon the number of logic gates that should be interconnected to perform the same input/output behaviour. This measure is more accurate that the simple number of gates [7], [15].

When the input to an electronic gate changes, there is a finite time delay before the change in input is seen at the output terminal. This is called the propagation delay of the gate and it differs from one gate to another. Of primary concern is the path from input to output with the highest total propagation delay. We estimate the performance of a given circuit using the worst-case delay path. The number of gate equivalent and an average propagation delay for each kind of gate are given in Table 5.4. The data were taken form [6].

Let C be a digital circuit that uses a subset (or the complete set) of the gates given in Table 5.4. Let $Gates(C)$ be a function that returns the set of all gates of circuit C and $Levels(C)$ be a function that returns the set of all the gates of C grouped by level. Notice that the number of levels of a circuit coincides with the cardinality of the set expected from function Levels. On the other hand, let $Val(X)$ be the Boolean value that the considered circuit C propagates for the input Boolean vector X assuming that the size of X coincides with the number of input signal required for circuit C. The fitness function, which allows us to determine how much an evolved circuit adheres to the specified constraints, is given as follows, wherein X represents the input values of the input signals while Y represents the expected output values of the output signals of circuit C, n denotes the number of output signals that circuit C has, function $Delay$ returns the propagation delay of a given gate as shown in Table 5.4 and Ω_1 and Ω_2 are the weighting coefficients [8] that allow us to consider both area and response time to evaluate the performance of an evolved circuit, with $\Omega_1 + \Omega_2 = 1$. Note that for each output signal error, the fitness function of Equation 5.3 sums up a penalty ψ. For implementation issue, we minimize the fitness function below for different values of Ω_1 and Ω_2.

$$\begin{aligned} Fitness(C) = & \sum_{i=0}^{n} \left(\sum_{i \mid Val(X_i) \neq Y_{i,j}} \psi \right) \\ & + \Omega_1 \sum_{g \in Gates(C)} GateEquiv(g) \\ & + \Omega_2 \sum_{l \in Levels(C)} \max_{g \in l} Delay(g) \end{aligned} \tag{5.3}$$

5.7 Comparative Results

In this section, we compare the evolved circuits to those obtained using the traditional methods, i.e. transition and Karnaugh maps. This is done for three different state machines that are generally used as benchmarks. These state machines are commonly called *shiftreg*, *lion*9 and *train*11. The detailed descriptions of these state machines can be found in [3]. The state assignments used are the best ones found so far. They also are the result of an evolutionary computation [14]. Theses state assignment are given in Table 5.2.

For each of these state machines, we evolved a minimal circuit that implements the required behaviour and compared it to the one engineered using the traditional method. Table 5.6 shows the details of this comparison. The schematics of the evolved circuit of state machines *shiftreg* are given in Fig. 5.14 and Fig. 5.15.

Table 5.6. Comparison of the traditional method vs. genetic programming

State machine	Number of gate-Equivalent		Response time	
	Traditional	GP	Traditional	GP
Shiftreg	30	12	0.85	0.423
Lion9	102	33	2.513	0.9185
Train11	153	39	2.945	0.8665

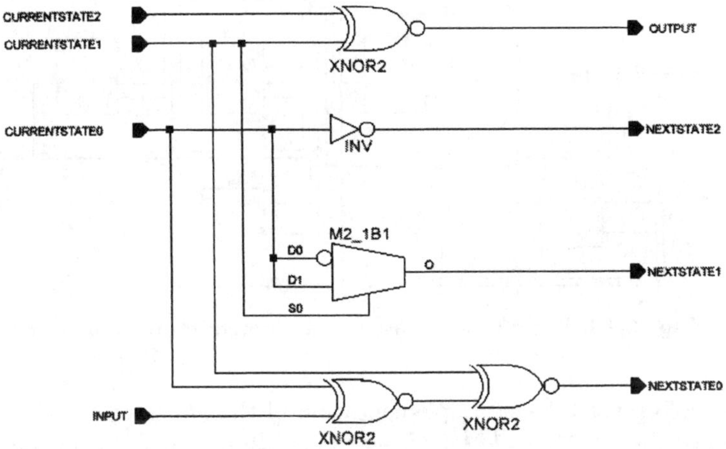

Fig. 5.14. First evolved control logic for state machine *shiftreg*

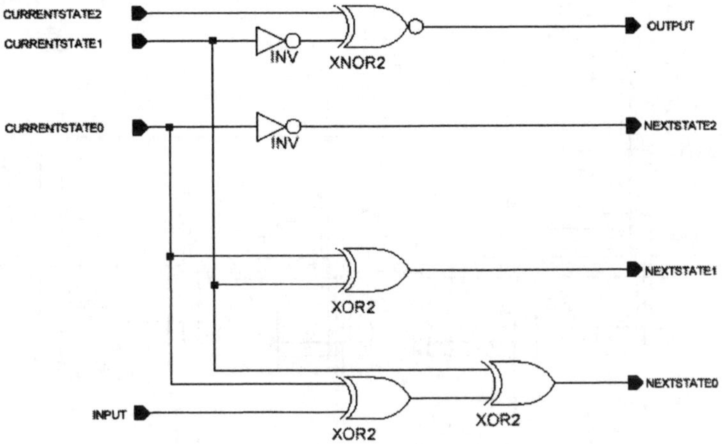

Fig. 5.15. Second evolved control logic for state machine *shiftreg*

The lookup table-based implementations of the *shiftreg* state machine for both control logics (i.e. of Fig. 5.14 and Fig. 5.15) exploits two 2-input, one 3-input and one 4-input lookup tables. The schematics are given in Fig. 5.10.

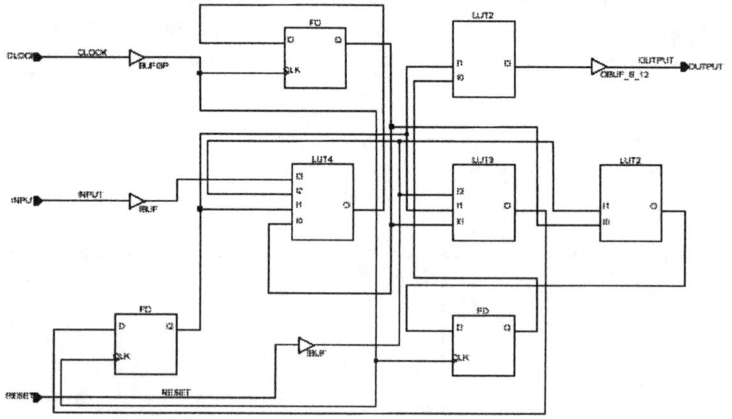

Fig. 5.16. Lookup table-based evolved architeture of *shiftreg*

The lookup table-based implementation of the *shiftreg* state machine as synthesised by the Xilinx^{TM} [17] uses four 2-input, one 3-input and one 4-input lookup tables. The schematics are given in Fig. 5.16.

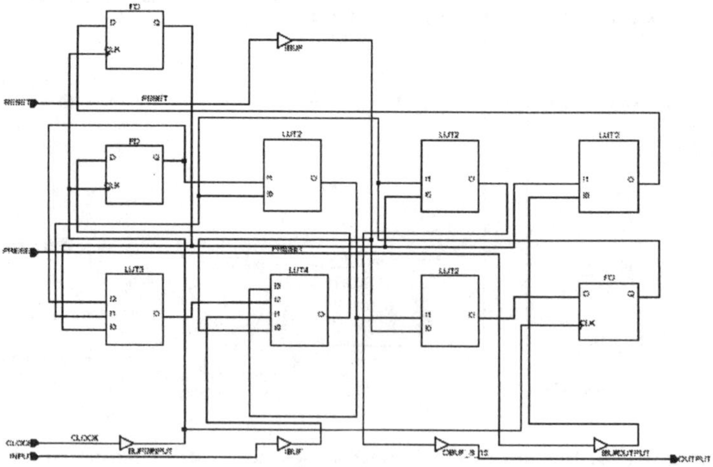

Fig. 5.17. Lookup table-based architeture of *shiftreg* as synthesised by Xilinx^{TM}

Fig. 5.18 and Fig. 5.19 show the evolved circuits for state machines *lion*9 and *train*11 respectively. It is clear that the evolved circuits are much better that those yield by the traditional methods in both terms hardware area and signal propagation delay.

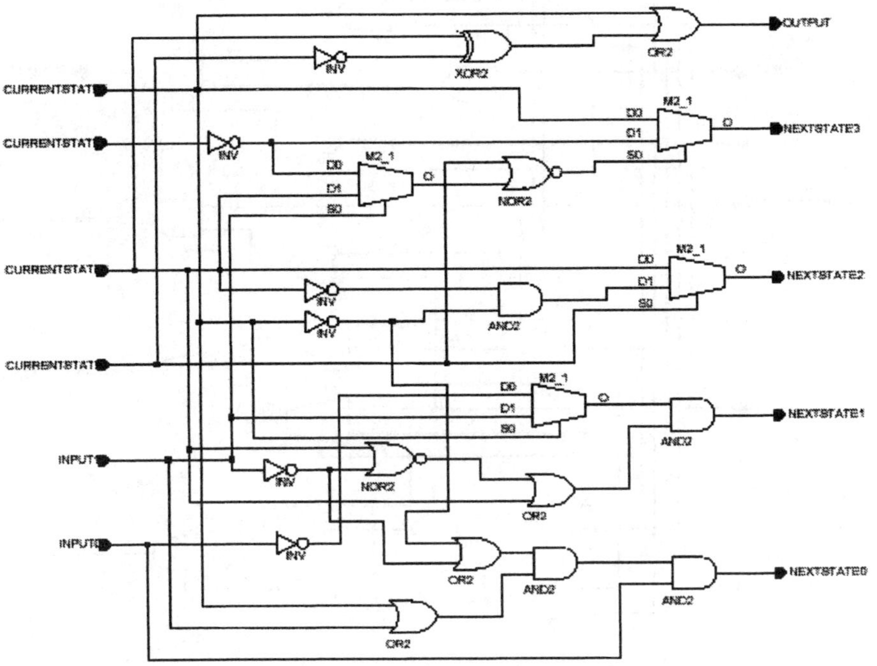

Fig. 5.18. The evolved control logic for state machine *lion*9

5.8 Summary

In this chapter, is divided into two main parts. In the first part, we exploited evolutionary computation to solve the NP-complete problem of state encoding in the design process of asynchronous finite state machines. We compared the state assignment evolved by our genetic algorithm for machine of different sizes evolved to existing systems. Our genetic algorithm always obtains better assignments. In the second part, we exploited genetic programming to synthesise the control logic used in asynchronous finite state machines. We compared the circuits evolved by our genetic programming-based synthesiser with that that would use the traditional method, i.e. using Karnaugh maps and transition maps. The state machine used as benchmarks are well known

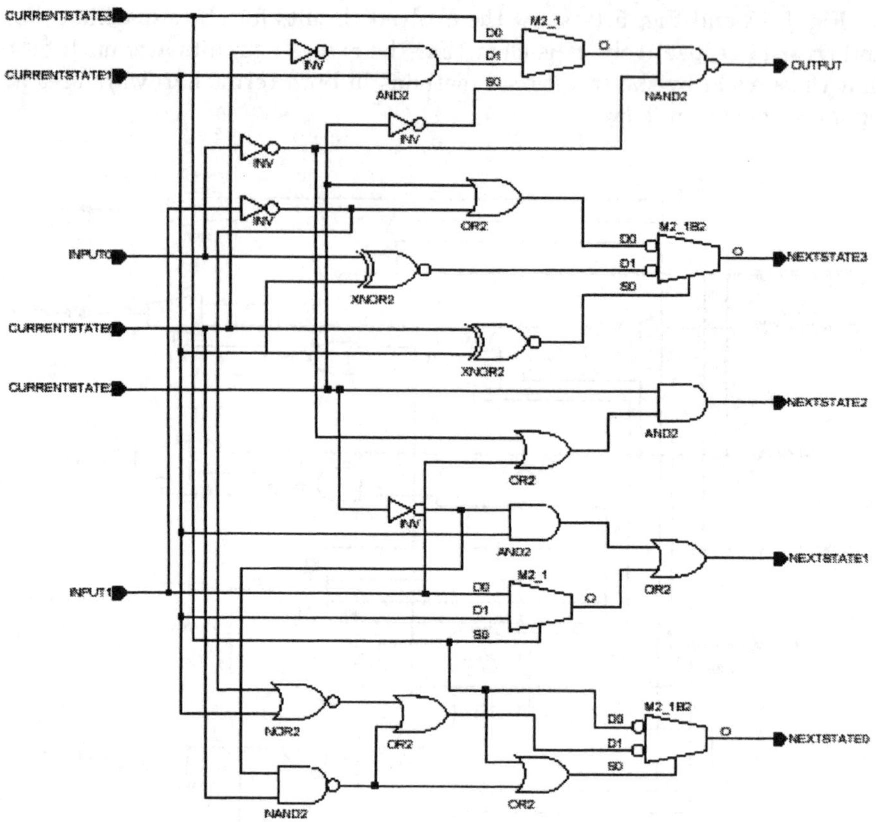

Fig. 5.19. The evolved control logic for state machine *train*11

and of different sizes. Our evolutionary synthesiser always obtains better control logic both in terms of hardware area required to implement the circuit and response time.

References

1. Amaral, J.N., Tumer, K. and Gosh, J., Designing genetic algorithms for the State Assignment problem, IEEE Transactions on Systems Man and Cybernetics, vol., no. 1999.
2. Armstrong, D.B., A programmed algorithm for assigning internal codes to sequential machines, IRE Transactions on Electronic Computers, EC 11, no. 4, pp. 466-472, August 1962.
3. Collaborative Benchmarking Laboratory, North Carolina State University, `http://www.cbl.ncsu.edu/pub/ Benchmark_dirs/LGSynth89/fsmexamples`, November 27th. 2003.

4. DeJong, K. and Spears, W.M., Using genetic algorithms to solve NP-complete problems, Proceedings of the Third International Conference on Genetic Algorithms, pp. 124-132, Morgan Kaufmann, 1989.
5. DeJong, K. and Spears, W.M., An analysis of the interacting roles of the population size and crossover type in genetic algorithms, In Parallel problem solving from nature, pp. 38-47, Springer-Verlag, 1990.
6. Davis, L., Handbook of Genetic Algorithms, Van Nostrand Reinhold, New York, 1991.
7. Ercegovac, M. D., Lang, T. and Moreno, J.H., Introduction to digital systems, John Wiley, 1999.
8. Fonseca, C.M. and Fleming, P.J., An overview of evolutionary algorithms in multi-objective optimization, Evolutionary Computation, 3(1):1-16.
9. Goldberg, D. E., Genetic Algorithms in Search, Optimisation and Machine Learning, Addison-Wesley, Massachusetts, Reading, MA, 1989.
10. Haupt, R.L. and Haupt, S.E., Practical genetic algorithms, John Wiley and Sons, 1998.
11. Humphrey, W.S., Switching circuits with computer applications, New York: McGraw-Hill, 1958.
12. Koza, J.R., Genetic Programming. MIT Press, 1992.
13. Michalewics, Z., Genetic algorithms + data structures = evolution program, Springer-Verlag, USA, third edition, 1996.
14. Nedjah, N. and Mourelle, L.M, Evolutionary state assignment for synchronous finite state machine, Proceedings of International Conference on Computational Science, Lecture Notes in Computer Science, Springer-Verlag, 2004.
15. Rhyne, V.T., Fundmentals of digital systems design, Computer Applications in Electrical Engineering Series, Prentice-Hall, 1973.
16. Villa, T. and Sangiovanni-Vincentelli, A. Nova: state assignment of finite state machine for optimal two-level logic implementation, IEEE Transactions on Computer-Aided Design, vol. 9, pp. 905-924, September 1990.
17. Xilinx, Project Manager, ISE 6.1i, `http://www.xlinx.com`.

6

Automating the Hierarchical Synthesis of MEMS Using Evolutionary Approaches

Zhun Fan[1,2], Jiachuan Wang[3], Kisung Seo[1], Jianjun Hu[4], Ronald Rosenberg[5], Janis Terpenny[3], and Erik Goodman[1]

[1] Department of Electrical and Computer Engineering, Michigan State University, East Lansing, MI, 48823, USA (fanzhun | ksseo | goodman)@egr.msu.edu, http://www.egr.msu.edu
[2] Department of Mechanical Engineering, Technical University of Denmark, DK-2800 Kgs. Lynby, Denmark
[3] Department of Industrial Engineering and Operations Research University of Massachusetts Amherst, Amherst, MA 01003 USA (jiacwang | terpenny)@ecs.umass.edu, http://www-unix.ecs.umass.edu
[4] Department of Computer Science and Engineering, Michigan State University, East Lansing, MI, 48823, USA hujianju@egr.msu.edu, http://www.egr.msu.edu/ hujianju
[5] Department of Mechanical Engineering, Michigan State University, East Lansing, MI, 48823, USA ron@egr.msu.edu, http://www.egr.msu.edu/ rosenber/

The complex device, component, and system design issues involved in integrated MEMS design call for a structured design methodology that borrows from VLSI design. In this chapter, we first discuss the hierarchy that is involved in a typical MEMS design. Then we move on to discuss how evolutionary approaches can be used to automate the hierarchical design and synthesis process for MEMS. At the system level, genetic programming, as a strong search tool, is used to generate and search in the topologically open-ended design space. Meanwhile, bond graphs are used to represent the lumped parameter models of MEMS that cut across mixed energy domains. The approach combining bond graphs and genetic programming can lead to satisfactory design candidates of system level models that meet the predefined behavioral specifications for designers to tradeoff. Then at the second level, namely the physical layout synthesis level, the selection of geometric parameters for component devices is formulated as a constrained optimization problem and addressed using a constrained GA approach. Considerations of feature size constraints can be incorporated into this approach very conveniently. A multiple-resonator microsystem design is taken as an example to illustrate the integrated design automation idea using evolutionary approaches at multiple levels.

6.1 Introduction

MicroElectroMechanical Systems (MEMS) is a rapidly expanding technology that offers new ways of combining sensing, actuation, signal processing, computing and communication functions on a miniature scale. Although MEMS is a promising technology, it is very surprising that we have only seen a handful of successful commercial MEMS products which the market has demanded in large quantities, including automotive accelerometers and gyroscopes, pressure sensors, ink-jet print heads and a few others. Prevalence of design and fabrication of MEMS application-specific integrated circuits (ASICS) analogous to electronic ASICS is still not seen. Due to the complexity and intricacy involved in MEMS design, designing MEMS still remains an art in most applications, requiring a large amount of investment of human resources, time and money. Much of the investment is consumed in the iterative trial-and-error design process. Automated design synthesis helps engineers to develop rapid, optimal configurations for a given set of performance and constraint guidelines, and thus to shorten typical development cycles for MEMS (with a given fabrication technology) by a large factor and to enable design of far more complex MEMS than can be handled today. Electronic Design Automation (EDA) has achieved great success in both industry and academia. However, analogous research in design automation for MEMS seems to lag far behind, although considering the close affinity of MEMS and VLSI - MEMS actually evolved from microelectronics and inherited the fabrication techniques of VLSI - the potential successful applications of design automation of MEMS appear to be promising. It turns out that translating the key insights of silicon evolution success into MEMS technologies is a much more challenging task than most people have expected. Major research topics to be addressed include: 1) developing a broad base of building blocks in MEMS technologies so that huge networks of micro-devices could be assembled into arbitrary architectures with desirable functionalities, 2) abstracting design hierarchies to stratify and conquer design complexity, thus making the design more amenable to an automated process, 3) improving models of computation and extending current synthesis methodologies to facilitate generation of viable design candidates and smoother transitions from conceptual and embodied designs to process fabrication. 4) combining MEMS component layout extraction and lumped-parameter bond graph simulation and design synthesis to provide MEMS designers with VLSI-like environments enabling faster design cycles and improved design productivity.

This chapter seeks to partially address the above challenges, especially the first two. The proposed hierarchical and evolutionary design framework for MEMS aims to eliminate tedious and repetitive design tasks, facilitate hierarchical problem decomposition, and combine the power of multiple evolutionary computation algorithms working simultaneously to identify better product designs and process solutions. In particular, we divide design representations of MEMS design into two levels, the system-level behavioral macromodel and the detailed-level physical geometric layout model. At the system level, we use a combination of genetic programming and bond graphs to automatically generate and search for viable design candidates represented by behavioral macromodels satisfying high-level design specifications. At the second detailed (layout) level, multiobjective constrained genetic algorithms are used to optimize the geometric parameters that relate the physical device model to the behavioral macromodel and meet more detailed design objectives.

6.2 Hierarchical MEMS Design Methodology

MEMS holds the promise of being amenable to structured automated design due to its similarities with VLSI. However, design and analysis of MEMS is much more complicated due to their multi-domain and intrinsically three-dimensional nature. In addition, because of limitations of fabrication technology, there are many constraints in design of MEMS. In MEMS, there are a number of levels of designs that need to be synthesized [1]. Usually the design process starts with basic capture of the schematic of the overall system, and then goes on through layout and construction of a 3-D solid model. So the first design level is the system level, which includes selection and configuration of a repertoire of planar devices or subsystems. The second level is 2-D layout of basic structures like beams to form the elementary planar devices. In some cases, if the MEMS is basically a result of a surface-micro machining process and no significant 3-D features are present, design of this level will end one cycle of design. More generally, modeling and analysis of a 3-D solid model for MEMS is necessary. However, even if we have obtained an optimized 3-D device shape, it is still very difficult to produce a proper mask layout and correct fabrication procedures. Automated mask layout and process synthesis tools would be very helpful to relieve designers from considering the fabrication details and focus on the functional design of the device and system [2]. After a "top-down" design path, a "bottom-up" verification process is usually followed to guarantee that at each design level the design specifications are met exactly as defined in Fig. 6.1. The ultimate goal is to develop tools for MEMS design to ensure first-pass success by having a well-defined "top-down" design path and "bottom-up" verification path.

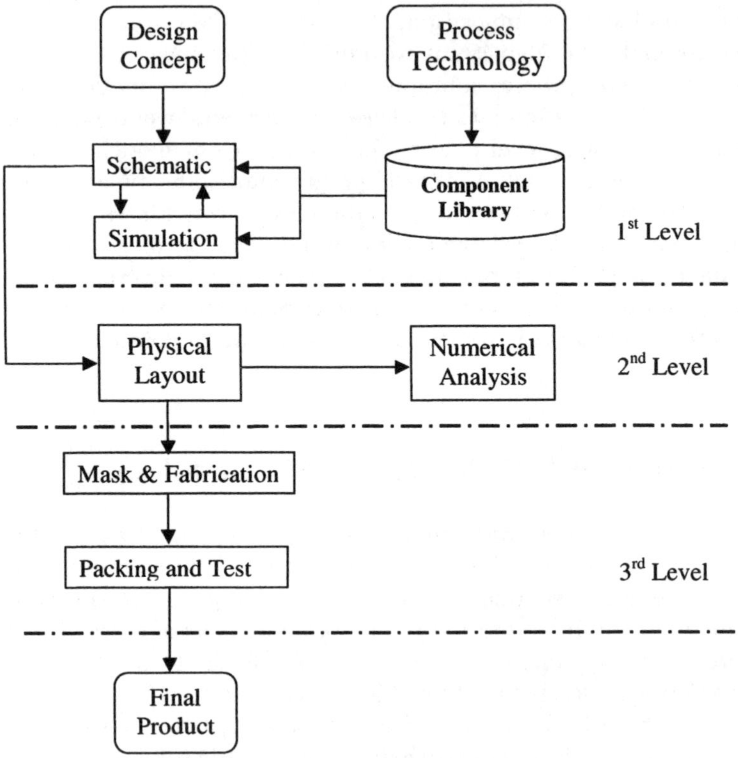

Fig. 6.1. Hierarchical design of MEMS

6.3 System-Level Synthesis of MEMS Using Genetic Programming and Bond Graphs

For system-level design, hand calculation is still the most popular method in current design practice. This is mainly because no powerful and widely accepted synthesis approach exists to automated design of multi-domain systems. In addition, most MEMS system-level design is accomplished by modeling entire microelectromechanical system as single behavioral entities having no lower hierarchical level in design. If there is any change in geometric parameters or topology, a whole new model must be created, and this substantially lengthens design cycles. Over the past two decades, computational design algorithms based on Darwin's principles of evolution have developed from academic curiosities into practical and effective tools for scientists and engineers. Gero, for example, investigates evolutionary systems as computational models of creative design and studies the relationships among genetic engineering, style emergence, and complex evolution [3]. Goodman et al. [4] studied evolution of engineering artifacts using heterogeneous parallel genetic algorithms. Koza has applied genetic programming to evolve analog filter circuits and can optimize the topology and sizing parameters of the evolved circuits simultaneously [5]. In this research, we use genetic programming as a strong search tool to explore the topologically open-ended design space for system-level behavioral models of MEMS. We also use bond graphs as a modeling tool to unify representations of mixed energy domains of MEMS. We call the overall approach the BG/GP approach.

6.3.1 Bond graphs

The reason we used bond graphs in research on MEMS synthesis is because MEMS are intrinsically multi-domain systems, unlike electronic systems. We need a uniform representation of MEMS so that designers can not only shift among different hierarchies of design abstractions but also can move around design partitions with different physical domains without difficulty. The bond graph is a modeling tool that provides a unified approach to the modeling and analysis of dynamic systems, especially hybrid multi-domain systems including mechanical, electrical, pneumatic, hydraulic components, etc. It is the explicit representation of model topology that makes the bond graphs a good candidate for use in open-ended design search. Fig. 6.2 shows an example of unique bond graphs representation of a resonator unit in three different application domains. It is also very natural to use bond graphs to represent a dynamic system, such as a mechatronic system, with cross-disciplinary physical domains and even controller subsystems (Fig.6.3). For notation details and methods of system analysis related to the bond graph representation, see [6]. Shah [7] identifies the importance of bond graphs for unifying multi-level design of multi-domain systems. Tay et al. [8] use bond graphs and GA to generate and analyze dynamic system designs automatically. This approach adopts a variational design method, which means they make a complete bond graph model first, and then change the bond graph topologically using a GA, yielding new design alternatives. However, the efficiency of this approach is hampered by the weak ability of GA to search in both topology and parameter spaces simultaneously. Terpenny and Jiachuan Wang have begun to explore combination of bond graphs and evolutionary computation [9]. Campell [10] also uses the idea of both bond graphs and genetic algorithms in his A-Design framework. In this research, we use an approach combining genetic programming and bond graphs to automate the process of design of dynamic systems to a significant degree.

6.3.2 Combining bond graphs and genetic programming

The most common form of genetic programming [5] uses trees to represent the entities to be evolved. Defining of a proper function set is one of the most significant steps in using genetic programming. It may affect both the search efficiency and validity of evolved results and is closely related to the selection of building blocks for the system being designed. By executing the genotype,

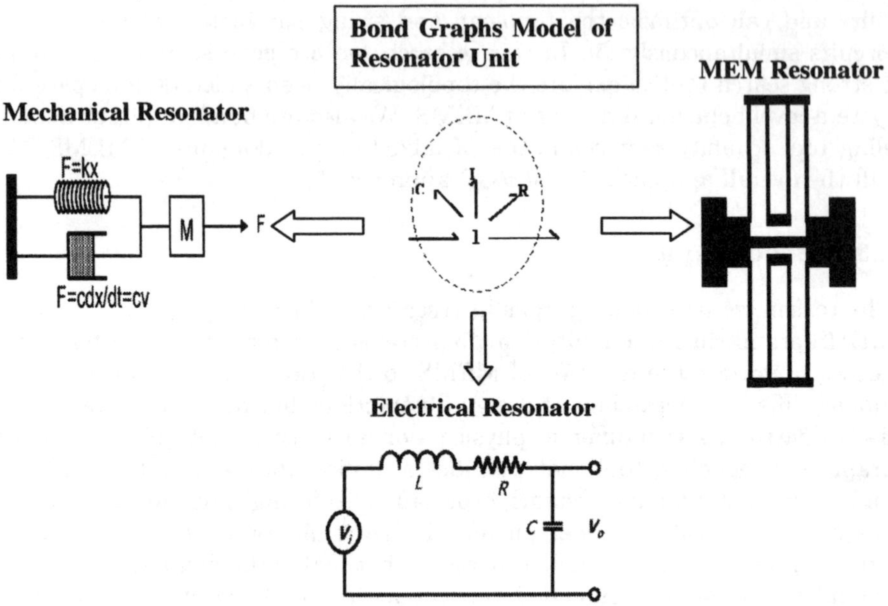

Fig. 6.2. Bond graphs representing a mechatronic system with mixed energy domains and a controller subsystem

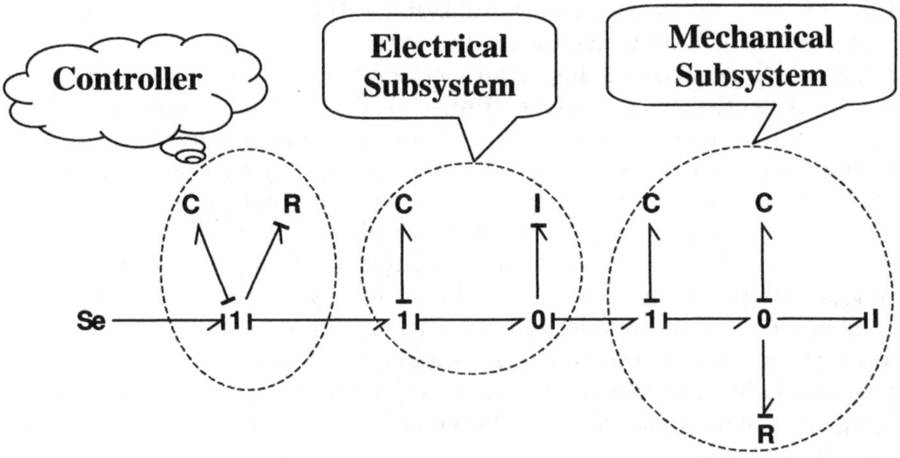

Fig. 6.3. One bond graph represents resonators in different application domains

a genetic programming tree that composes of functions in the function set as nodes of the tree, an arbitrary representative topology, or phenotype can be generated in a developmental manner. In this research, we have an additional dimension of flexibility in generating phenotypes, because bond graphs are used as modeling representations for multi-domain systems, serving as an intermediate representation between the mapping of genotype and phenotype, and can be interpreted as systems in different physical domains, chosen as appropriate to given circumstances. Fig. 6.4 illustrates the role of bond graphs in the mappings from genotypes to phenotypes. [11]

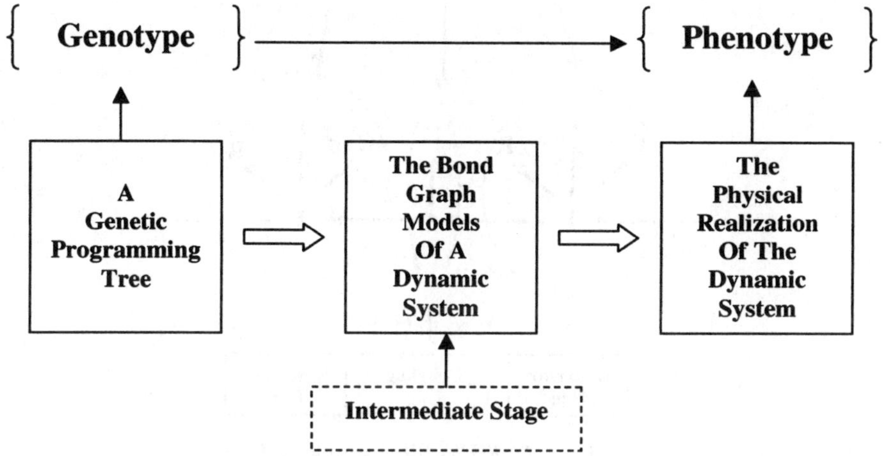

Fig. 6.4. Genotype-phenotype mapping

6.3.3 Filter topology

Automated synthesis of an RF MEM device, a micro-mechanical bandpass filter, is used as an example in this chapter [12]. Through analyzing two popular topologies used in surface micromachining of micro-mechanical filters, we found that they are topologically composed of a series of concatenated Resonator Units (RUs) and Bridging Units (BUs) or RUs and Coupling Units (CUs). Fig. 6.5 and Fig. 6.6 illustrates the layouts and bond graph representations of two widely accepted filter topologies I and II [12]. Their corresponding bond graph representations are also shown.

6.3.4 Function set

In this research, a GP function set is presented and listed in Table 6.1. Examples of operators, namely insert-CU and insert-RU, are illustrated in Fig.s 6.7 and 6.8. Fig. 6.7 explains how the insert-CU function works. A Coupling

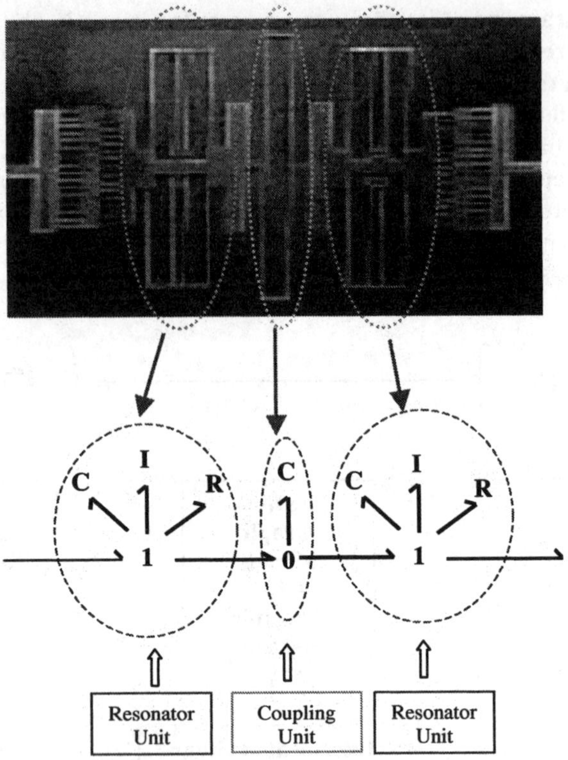

Fig. 6.5. MEM filter topology I

Unit (CU) is a subsystem that is composed of a capacitor attached with a 0-junction in the center and two bonds connecting 1-junctions at the left and right ends. After execution of the insert-CU function, an additional modifiable site (2) appears at the rightmost newly created bond. As illustrated in Fig. 6.8, a resonator unit (RU), composed of one I, R, and C component all attached to a 1-junction, is inserted in an original bond with a modifiable site through the insert-RU function. After the insert-RU function is executed, a new RU is created and one additional modifiable site, namely bond (3), appears in the resulting phenotype bond graph, along with the original modifiable site bond (1). The newly-added 1-junction also has an additional modifiable site (2). As components C, I, and R all have parameters to be evolved, the insert-RU function has three corresponding ERC-typed sites, (4), (5), and (6), for numerical evolution of parameters.

6.3.5 Design embryo

All individual genetic programming trees create bond graphs from an embryo. Selection of the embryo is also an important topic in system design, especially

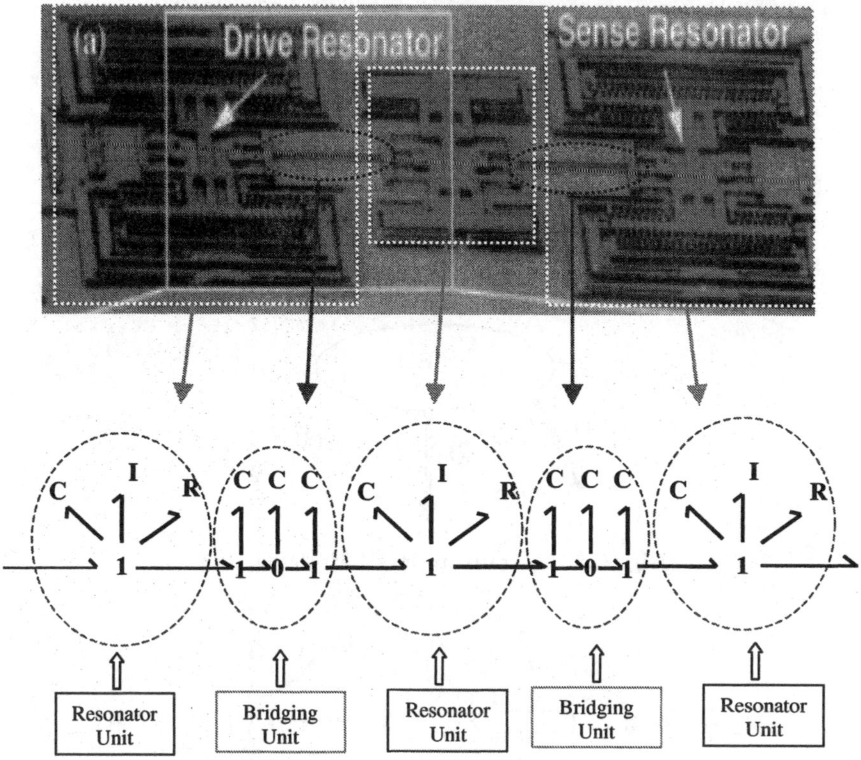

Fig. 6.6. MEM filter topology II

Table 6.1. Operators in modular function set

Operator Name	Functionality
Insert-RU	insert a resonator unit
Insert-CU	insert a coupling unit
Insert-BU	insert a bridging unit
Add-RU	add a resonator unit
Insert-J01	insert a 0-1-junction
Insert-CIR	insert a special CIR component
Insert-CR	insert a special CR component

for multi-port systems. In our filter design problems, we use the bond graph shown in Fig. 6.9 as our embryo.

6.3.6 Fitness function

Within the frequency range of interest, $f_{rang} = [f_{min}, f_{max}]$, uniformly sample 100 points. Here, $f_{rang} = [0.1, 1000\text{K}]$ Hz. Compare the magnitudes of the

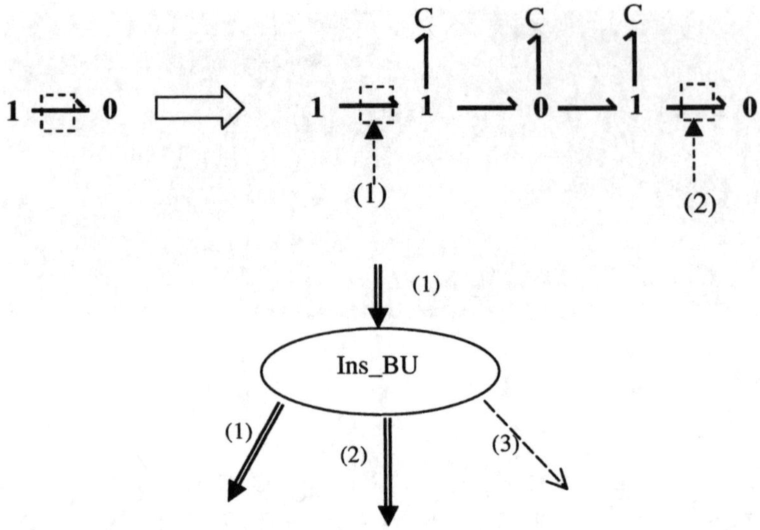

Fig. 6.7. Operator to insert Bridging Unit

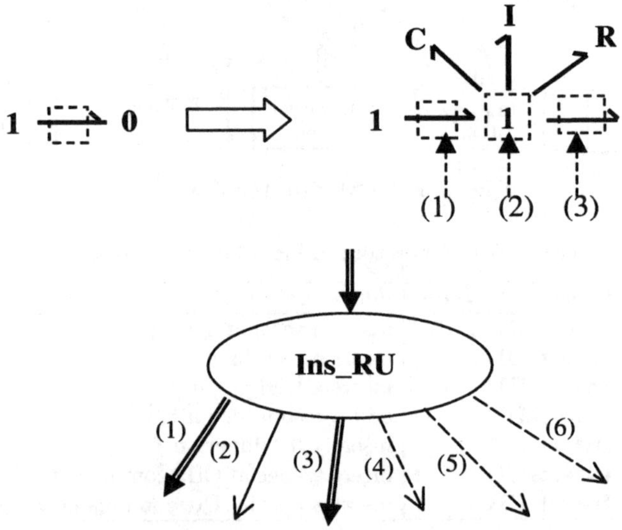

Fig. 6.8. Operator to insert Resonator Unit

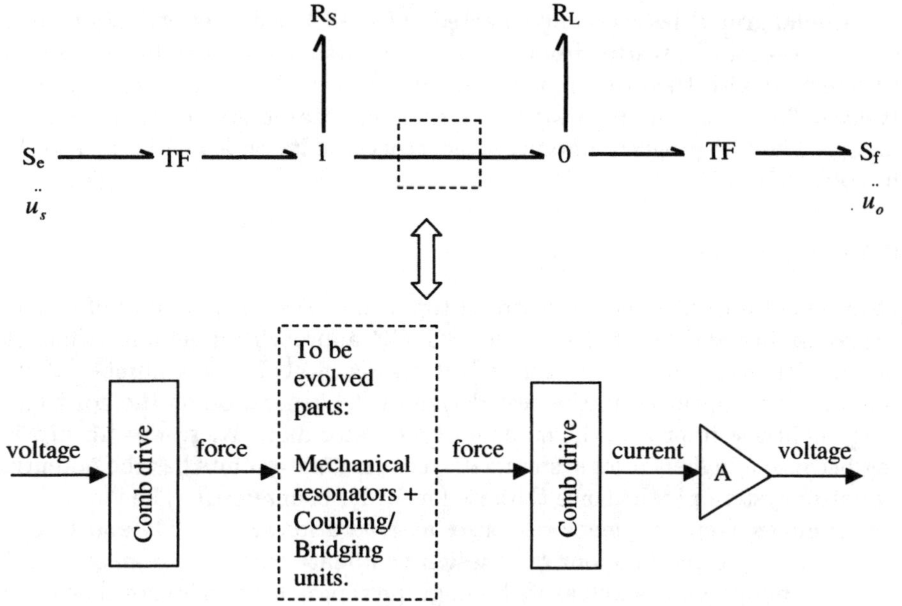

Fig. 6.9. Design Embryo of a Micro-Electro-Mechanical Filter

frequency response at target magnitudes, which are 1.0 within the pass frequency range of [316, 1000] Hz, and 0.0 otherwise, between 0.1 and 1000KHz.

6.3.7 Experimental setup

Three major code modules were created in this work. The algorithm kernel of HFC-GP was a strongly typed version [13] of an open software package developed in our research group – lilgp. Parameters for lilgp are shown in the tableau 6.2.

Table 6.2. Parameter settings for genetic programming

Parameter	Setting
population size	500 in each of thirteen subpopulations
initial population	half and half
initial depth	4-6
maximum depth	50
maximum nodes	5000
selection method	tournament with size 7
crossover rate	0.9
mutation rate	0.3

A bond graph class was implemented in C++. The fitness evaluation package is C++ code converted from Matlab code, with hand-coded functions used to interface with the other modules of the project. The commercial software package 20Sim was used to verify the dynamic characteristics of the evolved design. The GP program obtains satisfactory results on a Pentium-IV 1GHz in 1000 1250 minutes.

6.3.8 Experimental result

Experimental results show the strong topological search capability of genetic programming and feasibility of our BG/GP approach for finding realizable designs for micro-mechanical filters [14]. In Fig. 6.11, K is the number of resonator units appearing in the best design of the generation on the horizontal axis. As fitness improves, the number of resonator units, K, grows - unsurprising because a higher-order system with more resonator units has the potential of better system performance than its low-order counterpart. The plot of corresponding system frequency responses at generations 27, 52, 117 and 183 are shown in Fig. 6.10. A layout of a design candidate with four resonators and three coupling units as well as its bond graph representation is shown below in Fig. 6.12. Notice that the geometry of resonators may not show the real sizes and shapes of a physical resonator and the layout figure only serves as a topological illustration. Using the BG/GP approach, it is also possible to explore novel topologies of MEM filter design. In this case, we may not necessarily use a strictly realizable function set. Instead, a semi-realizable function set may be used to relax the topological constraints, with the purpose of finding new topologies not realized before but still realizable after careful design. Fig. 6.13 gives an example of a novel topology for a MEM filter design. An attempt to fabricate this kind of topology is being carried out in a university research setting.

6.4 Second-Level Physical Layout Synthesis Formatting the Headings

Layout synthesis automatically generates valid or optimized geometric sizing parameters for cell components, which in most cases are commonly used micromechanical devices with fixed topologies, according to engineering design objectives. In this research, the cell component is a resonator device in MEMS domain. The design objectives come from either high-level specifications such as behavioral model parameters that need to be satisfied, or from layout-level objectives such as minimum areas occupied. Our approach is to model the design problem as a formal constrained optimization problem, and then solve it with powerful optimization techniques, resulting in a tool that automates the design synthesis of MEMS structures. Two categories of optimization techniques are used: one category includes stochastic algorithms

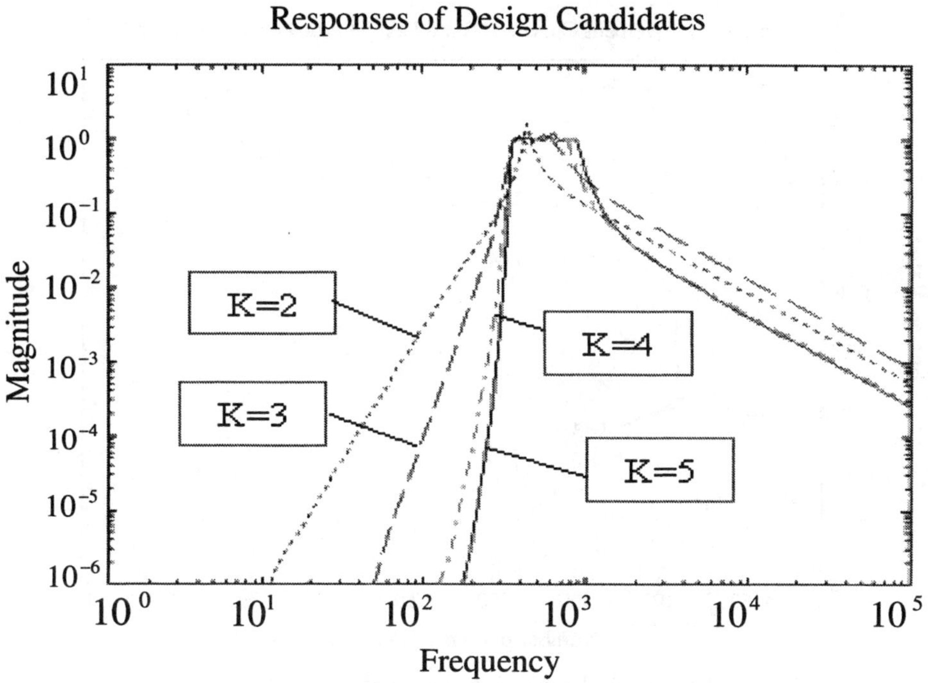

Fig. 6.10. Frequency responses of a sampling of design candidates, which evolved topologies with larger numbers, K, of resonators as the evolution progressed. All results are from one genetic programming run of the BG/GP approach

such as genetic algorithms, and the other category includes deterministic algorithms such as nonlinear programming. For both categories, the process of solving the optimization problem involves determining the design variables, the design constraints, and the design objective. We decided to use 14 design variables for an example cell component, a folded-flexure comb-drive microresonator fabricated in a polysilicon surface microstructural process (Fig. 6.14) in this research. Design variables and their constraints are listed as follows (Fig. 6.15) [15]:

It is noted that the first 13 design variables have units of μm. The fourteenth design variable has units of volts. In addition, we assume t = w_c = g = d. in our design for simplicity. Some design variables are predefined: they are $w_{ba} = 11$, $w_{ca} = 14$, $\delta = 4$, N = 10 . The constraints for the design variables are listed below.

$$2 \leq L_b \leq 400, 2 \leq w_b \leq 20, 2 \leq L_t \leq 400, 2 \leq w_t \leq 20 \tag{6.1}$$

$$2 \leq L_{sy} \leq 400, 10 \leq w_{sy} \leq 400, 10 \leq w_{sa} \leq 400, 10 \leq w_{cy} \leq 400 \tag{6.2}$$

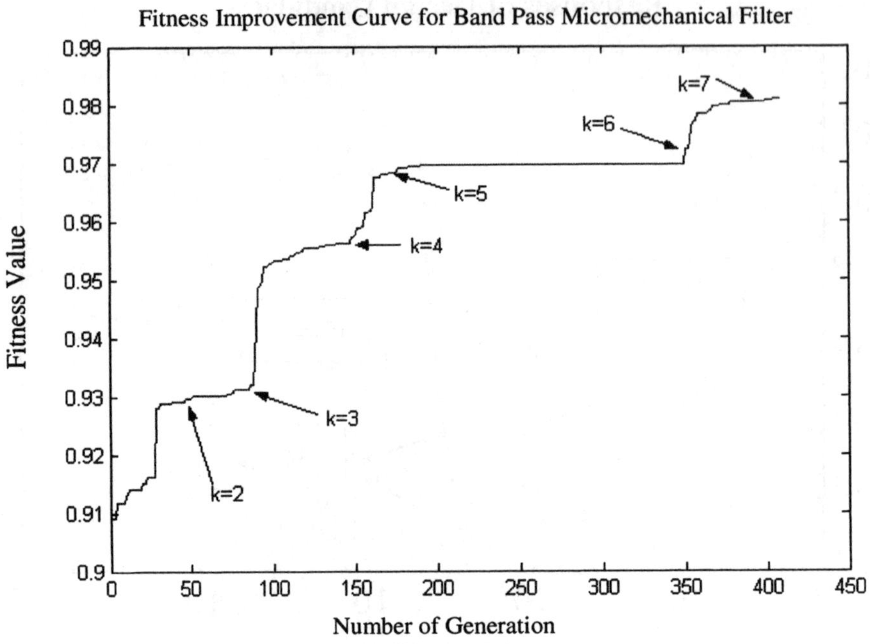

Fig. 6.11. Fitness improvement curve

$$10 \leq L_{cy} \leq 700, 8 \leq L_c \leq 400, 2 \leq w_c \leq 20, 2 \leq L_{sa} \leq 400 \quad (6.3)$$

$$4 \leq x_o \leq 400, 0 \leq V \leq 100 \quad (6.4)$$

also a number of design constraints for the microresonator cell component, including both geometric constraints and functional constraints. In this chapter, without loss of generality, we consider the following constraints:

$$0 \leq L_{cy} + 2g + 2w_c \leq 700 \quad (6.5)$$

$$0 \leq L_{sy} + 2L_b + 2w_t \leq 700 \quad (6.6)$$

$$0 \leq 3L_t + w_{sy} + 4L_c - 2x_0 + 2w_{cy} + 2w_{ca} \leq 700 \quad (6.7)$$

$$4 \leq L_c - (x_0 + x_{disp}) \leq 200 \quad (6.8)$$

Among them, the first three are linear constraints, and the fourth is a nonlinear constraint because the term x_{disp} is highly nonlinear. $x_{disp} = QF_{e,x}/K_x$, where $F_{e,x} = 1.12\varepsilon_0 NV^2 t/g$,

Suppose that in the system-level synthesis, we get a set of behavioral parameters for the cell component of a microresonator as

$$K_x = 0.27N/m \quad (6.9)$$

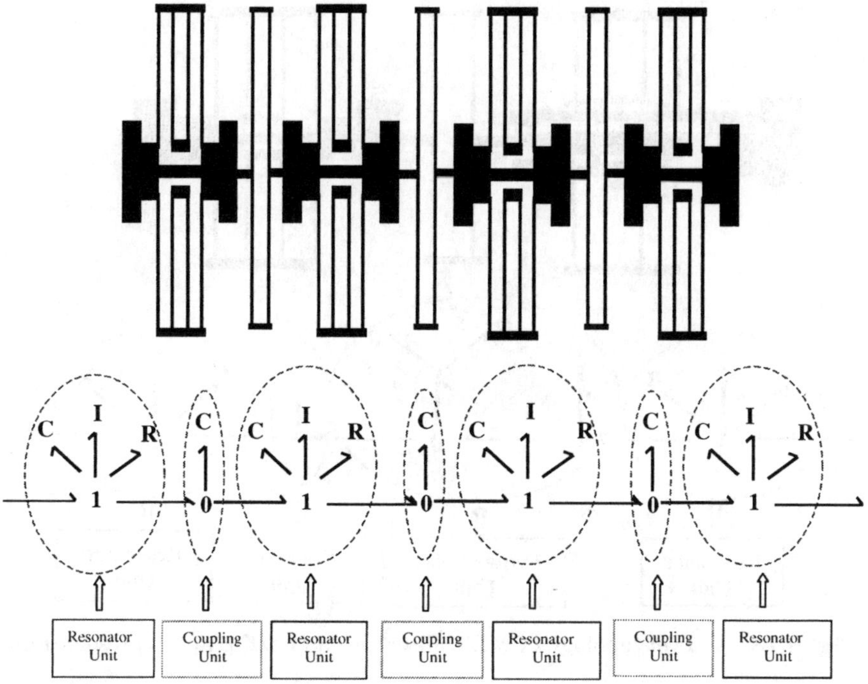

Fig. 6.12. Layout and bond graph representation of a design candidate from the experiment, with four resonator units coupled with three coupling units

$$B_x = 5.18 \times 10^{-6} kg \cdot m^2 \tag{6.10}$$

$$M_x = 4.0 \times 10^{-6} kg \tag{6.11}$$

Then we have three additional equation constraints.Equations to relate the design variables and the three behavioral model parameters are as follows:

$$K_x = \frac{2Etw_b^3}{L_b^3} \frac{L_t^2 + 14\alpha L_t L_b + 36\alpha^2 L_b^2}{4L_t^2 + 41\alpha L_t L_b + 36\alpha^2 L_b^2} \tag{6.12}$$

$$B_x = \mu[(A_s + 0.5A_t + 0.5A_b)(1/d + 1/\delta) + A_c/g] \tag{6.13}$$

$$M_x = M_s + \frac{1}{4}M_t + \frac{12}{35}M_b \tag{6.14}$$

where $\alpha = (W_t/W_b)^3$, $M_s = \rho A_s$, $M_t = \rho A_t$, $M_b = \rho A_b$, and

$$A_s = w_{sa}L_{sa} + 2w_{sy}L_{sy} \tag{6.15}$$

$$A_t = 2w_{ca}L_{cy} \tag{6.16}$$

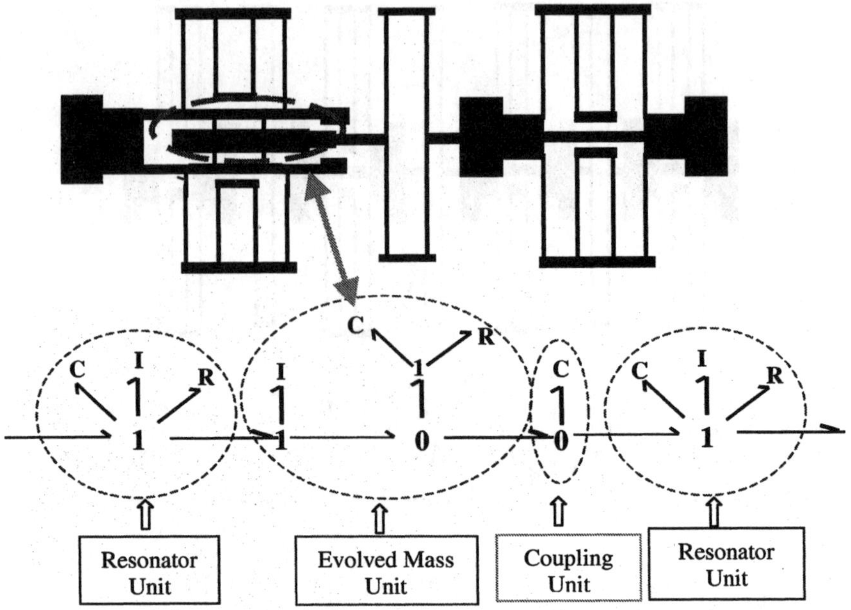

Fig. 6.13. A novel topology of MEM filter and its bond graph representation

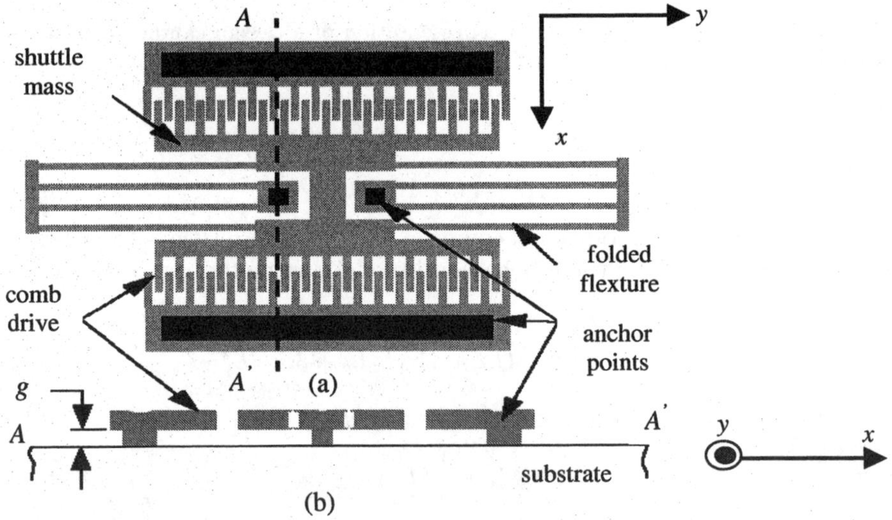

Fig. 6.14. A folded-flexure comb-drive microresonator fabricated in a polysilicon surface microstructural process a) Layout b) Cross-section A-A' [15]

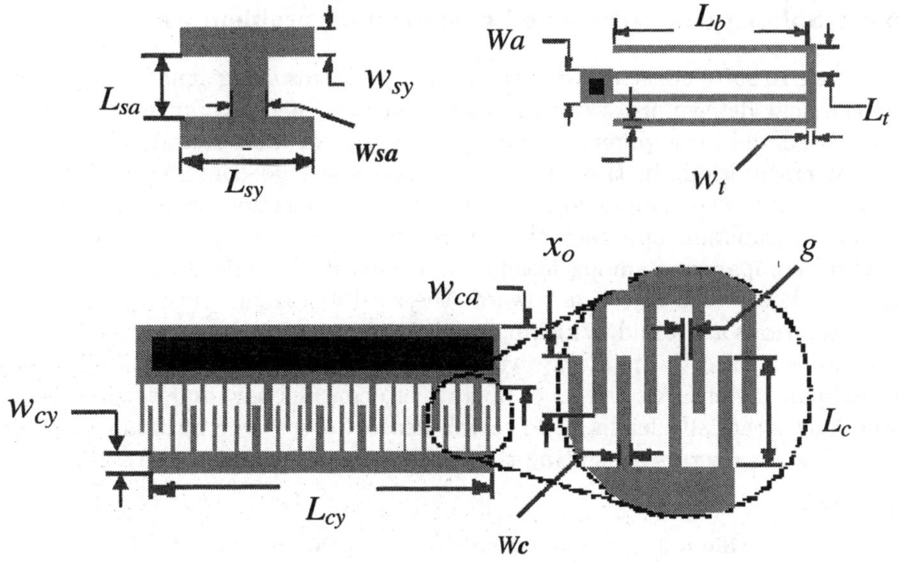

Fig. 6.15. Major design variables for microresonators

$$A_b = 8L_b w_b + 2w_t(2L_t + w_a + 2w_b) \tag{6.17}$$

As an alternative, we can also put reformulations of these three constraint equations into our design objectives, expressing them as differences to be minimized. In that case, we actually deal with a multi-objective constrained optimization problem.We take the objective function with the following normalized Sum of Squared Error (SSE) format:

$$\begin{aligned} f(\overrightarrow{x}) = &\tfrac{1}{0.27^2}(K_x - 0.27)^2 + \tfrac{1}{(5.18\times10^{-6})^2}(B_x - 5.18\times10^{-6})^2 \\ &+ \tfrac{1}{(4.0\times10^{-6})^2}(M_x - 4.0\times10^{-6})^2 \end{aligned} \tag{6.18}$$

Finally, it is important to note the role of feature size in VLSI and MEMS design. Feature size, which is often represented as λ , means the minimum size a particular design can achieve, based on specific fabrication procedures. In addition, the actual sizes of geometric shapes should be integer multiples of the feature size λ, such as λ, 2λ , 5λ , 10λ etc. In this research, we set $\lambda = 0.09\mu\text{m}$.

While it is very difficult for many numerical optimization approaches (for example, gradient-based approaches) to include considerations of feature size constraints [15], it is quite convenient for genetic algorithms to do so. We need to modify the objective function only slightly, mapping real values of design variables to integer multiples of the feature size λ before using them in formulations of constraints and objectives. No modifications to the genetic algorithm are needed.

6.4.1 Solving the constrained optimization problem using GA

In trying to solve constrained optimization problems using genetic algorithms or classical deterministic optimization methods, penalty function methods have been the most popular approach, because of their simplicity and ease of implementation. In this chapter, we use a special constrained GA that exploits pair-wise comparisons in a tournament selection operator to devise a penalty function approach that does not require any penalty parameter. Careful comparisons among feasible and infeasible solutions are made so as to provide a search direction towards the feasible region. Once sufficient feasible solutions are found, a niching method (along with a controlled mutation operator) is used to maintain diversity among feasible solutions. This allows a real-parameter GA's crossover operator to continuously find better feasible solutions, gradually leading the search nearer to the true optimum solution [16]. The parameters for setting the constrained GA are listed in Table 6.3.

Table 6.3. The parameters for setting the constrained GA

Parameter	Setting
variable boundaries	rigid
population size	500
total number of generations	100
crossover probability	0.9
mutation probability	0.15
niching parameter	0.9
exponent(n for SBX)	2.0
exponent(n for mutation)	50.0

In nine runs of the genetic algorithm using different random seeds, we obtained the sizing parameters and values of the objective function NSSE (to be minimized) listed in Table 6.4.

It can be seen that during the nine runs using different seeds, the constrained GA performs very steadily. Almost all runs achieved NSSE within the range of 1.0E-06. The biggest NSSE is 1.4E-05. However, the normalized squared sum of errors of 1.4E-05 is still considered very good result. It also appears that there are many alternatives and rather different ways in which parameters can be set and still produce behavior rather close to that desired.

6.5 Summary

This chapter has suggested a design methodology for automatically synthesizing hierarchical designs for MEMS. While there has been much research using evolutionary computation techniques to synthesize MEMS [2][17], this is the

Table 6.4. Layout parameters obtained in nine GA runs(different random seeds)

Run No.	1	2	3	4	5	6	7	8	9
$L_b(\lambda m)$	261.6	261.4	261.1	262.4	262.3	260.8	261.7	261.9	262.6
$w_b(\lambda m)$	1.98	1.98	1.98	1.98	1.98	1.98	1.98	1.98	1.98
$L_t(\lambda m)$	3.87	4.32	3.87	3.60	8.46	2.43	2.52	5.13	6.84
$w_t(\lambda m)$	2.70	2.25	2.52	2.52	2.25	1.98	1.98	2.88	3.33
$L_{sy}(\lambda m)$	3.69	2.88	2.07	4.41	1.98	1.98	3.60	1.98	2.79
$w_{sy}(\lambda m)$	14.13	12.60	15.93	11.52	19.80	9.99	11.52	15.30	12.60
$w_{sa}(\lambda m)$	18.63	18.18	10.98	11.70	11.34	11.16	10.17	11.70	14.58
$w_{cy}(\lambda m)$	146.16	151.83	122.31	141.12	137.25	56.61	110.70	76.14	247.50
$L_{cy}(\lambda m)$	15.66	20.79	23.85	17.37	23.85	30.69	22.68	21.96	8.91
$L_c(\lambda m)$	199.3	187.3	174.1	202.4	181.9	154.7	188.2	162.1	161.9
$w_c(\lambda m)$	1.98	1.98	1.98	1.98	1.98	1.98	1.98	1.98	1.98
$L_{sa}(\lambda m)$	2.25	2.16	2.52	2.43	2.88	1.98	2.70	2.70	6.30
$x_o(\lambda m)$	10.26	96.12	24.66	34.92	10.35	14.94	30.87	20.34	25.83
$V(volt)$	66.06	70.29	75.51	64.98	72.27	85.14	69.93	81.09	81.27
NSSE	4.0E-6	3.0E-6	3.0E-6	1.0E-6	1.0E-6	1.4E-5	2.0E-6	2.0E-6	1.0E-6

first work reported to seek to automate the hierarchical MEMS synthesis process in an integrated framework. Our first step is to synthesize system-level behavioral models using a combination of genetic programming and bond graphs. Then as the second step, we use a constrained genetic algorithm to automatically optimize the geometric sizing parameters for the cell components. An example of MEM filter design with coupling of multiple microresonators is used to illustrate the approach. Extension of this work can lead to a composable design and synthesis environment for micromechatronic systems [18]. In addition, target cascading in optimal system design needs to be investigated in depth to propagate the desirable top-level design specifications to appropriate specifications for the various subsystems and components in a consistent and efficient manner [19][20]. More work is underway to improve the efficiency of genetic programming to explore topologically open-ended design spaces, and the robustness of the constrained genetic algorithm to solve real-world constrained optimization problems.

References

1. Fedder, G.K. and Q. Jing, A Hierarchical Circuit-Level Design Methodology for Microelectromechanical Systems, IEEE Transactions on Circuits and Systems II (TCAS), vol. 46, no. 10, pp. 1309-1315, Oct. 1999
2. Ma, L. and E. K. Antonsson, Automated Mask-Layout and Process Synthesis for MEMS, Technical Proceedings of the 2000 International Conference on Modeling and Simulation of Microsystems, pp. 20-23, 2000
3. Gero J. S., Computers and Creative Design, in M. Tan and R. Teh (eds), The Global Design Studio, National University of Singarpo pp. 11-19, 1996

4. Eby, D., R. Averill, B. Gelfand, W. Punch, O. Matthews, E. Goodman, An Injection Island GA for Flywheel Design Optimization. 5th European Congress on Intelligent Techniques and Soft Computing, EUFIT'97. Vol. 1. pp. 687-691, 1997
5. Koza J. R., Genetic Programming II: Automatic Discovery of Reusable Programs, MIT Press, 1994
6. Rosenberg R. C., Reflections on Engineering Systems and Bond Graphs, Trans. ASME J. Dynamic Systems, Measurements and Control, 115, pp. 242-251, 1993
7. Vargas-Hernandez N., J. Shah, Z. Lacroix, Development of a Computer-Aided Conceptual Design Tool for Complex Electromechanical Systems, Computational Synthesis: From Basic Building Blocks to High Level Functionality, Papers from the 2003 AAAI Symposium Technical Report SS-03-02 pp. 255-261, 2003
8. Tay E. H., Flowers W. and Barrus J., Automated Generation and Analysis of Dynamic System Designs, Research in Engineering Design 10: pp. 15-29, 1998
9. Wang, J. and Terpenny, J., Interactive Evolutionary Solution Synthesis in Fuzzy Set-based Preliminary Engineering Design, Special Issue on Soft Computing in Manufacturing, Journal of Intelligent Manufacturing, Vol. 14. pp. 153-167, 2003
10. Campbell, M., Cagan J. and Kotovsky K., Agent-based Synthesis of Electro-Mechanical Design Configurations, Journal of Mechanical Design, Vol. 122. No. 1, pp. 61-69, 2000
11. Fan Z., Hu J., Seo K., Goodman E., Rosenberg R., and Zhang B., Bond Graph Representation and GP for Automated Analog Filter Design, GECCO-2001 Late-Breaking Papers, San Francisco, pp. 81-86, 2001
12. Wang K. and Nguyen C. T. C., High-Order Medium Frequency Micromechanical Electronic Filters, Journal of Microelectromechanical Systems, pp. 534-556, 1999
13. Luke S., Strongly-Typed, Multithreaded C Genetic Programming Kernel, http://www.cs.umd.edu/users/-seanl/gp/patched-gp/, 1997
14. Fan Z., Seo K., Rosenberg R., Hu J., Goodman E., System-Level Synthesis of MEMS via Genetic Programming and Bond Graphs, Proc. 2003 Genetic and Evolutionary Computing Conference, Chicago, Springer, Lecture Notes in Computer Science, July,2058-2071, 2003
15. Fedder G. and Mukherjee T., Physical Design for Surface-Micromachined MEMS, Proceedings of the Fifth ACM/SIGDA Physical Design Workshop, April, pp. 53-60, 1996
16. Deb K., An efficient constraint handling method for genetic algorithms, Comput. Methods Appl. Mech. Engrg., Vol. 186, pp. 311-338, 2000
17. N. Zhou, B. Zhu, A.M. Agogino, K.S.J. Pister, Evolutionary Synthesis of MEMS (Microelectronic Mechanical Systems) Design, Proceedings of ANNIE 2001, Intelligent Engineering Systems through Artificial Neural Networks, Volume 11, ASME Press, pp. 197-202, 2001
18. C. J. J. Paredis, A. Diaz-Calderon, R. Sinha, and P. K. Khosla, Composable Models for Simulation-Based Design, Engineering with Computers 17, pp. 112-128, 2001
19. Kim, H.M., Michelena, N.F., Papalambros, P.Y., and Jiang, T., " Target Cascading in Optimal System Design, Proceedings of the 2000 ASME Design Automation Conference, DAC-14265, Baltimore, Maryland, USA, September 10-13, 2000

20. Kim, H.M., Target Cascading in Optimal System Design, Ph.D. Dissertation, Department of Mechanical Engineering, University of Michigan, Ann Arbor, Michigan, USA, 2001

7

An Evolutionary Approach to Multi-FPGAs System Synthesis

F. Fernández de Veja[1], J.I. Hidalgo[2], J.M. Sánchez[1], and J. Lanchares[3]

[1] Departamento de Informática, Centro Universitario de Mérida, Universidad de Extremadura, C/ Sta Teresa de Jornet, 38 - 06800 Mérida, Spain
fcofdez@unex.es, http://atc.unex.es/pacof

[2] Departmento de Arquitectura de Computadores y Automática, Facultad de Informática, Universidad Complutense de Madrid, C/ Juan del Rosal, 8 -28040 Madrid Spain
hidalgo@dacya.ucm.es, http://www.dacya.ucm.es/hidalgo

[3] Departamento de Informática, Escuela Politécnica, Universidad de Extremadura, Spain

[4] Departmento de Arquitectura de Computadores y Automática, Facultad de Informática, Universidad Complutense de Madrid, C/ Juan del Rosal, 8, Spain,
julandan@dacya.ucm.es

In this chapter we explain in detail a methodology for Multi-FPGA systems (MFS) design. MFSs are hardware platforms used for a great variety of applications, including dynamically re-configurable hardware applications, digital circuit emulation, and numerical computation. There are a lot of MFS not only academical, but also commercial implementations. We describe a set of techniques based on evolutionary algorithms (EA), and we show that they are capable of solving all of the design tasks (partitioning, placement and routing). Firstly a hybrid compact genetic algorithm (HcGA) solves the partitioning problem and then genetic programming (GP) is used to obtain a solution for the two remaining tasks.

7.1 Introduction

Field Programmable Gate Arrays (FPGAs) are integrated devices used on the implementation of digital circuits by means of a configuration or programming process. There are different manufacturers and several kind of FPGAs are available. We will focus on those called island-based FPGAs. This model includes three main components: configurable logic blocks, input-output blocks and connection blocks (see figure 7.1). Configurable logic blocks (CLBs) are used to implement all the logic circuitry. They are positioned in a matrix way in the device, and they have different configuration possibilities. Input-output blocks (IOBs) are responsible for connecting the circuit implemented by the CLBs with any external system. The third class of components are connection blocks (switch-boxes and interconnection lines). They are the elements available for the designer to make the internal routing of the circuit. In most occasions we need to use some of the CLBs to accomplish the routing [1].

When the size of an FPGA is not enough to implement large circuits, the designer must think on higher reconfigurable platforms, in other words, on the use of Multi-FPGA system (MFS) [2]. These systems can eventually include, in addition to several FPGA devices, memories and other hardware elements. MFS are used for dynamically re-configurable hardware applications [3][4], digital circuit emulation [5], numerical computation [6], etc [7] [8]. The two most widely used topologies are the mesh and crossbar types. Mesh MFSs have simple routing methodologies, an easy expandability, FPGAs are connected in the nearest-neighbor pattern, and all devices are used for the same functionality. Fig. 7.2 (a) represents a mesh-topology MFS. A Crossbar MFS model is depicted on figure 7.2 (b). On this style, FPGAs are separated into logic and routing chips. Crossbar distributions are normally designed for some specific problems, but they usually waste logic and routing resources. For these reasons we have focused on mesh topologies.

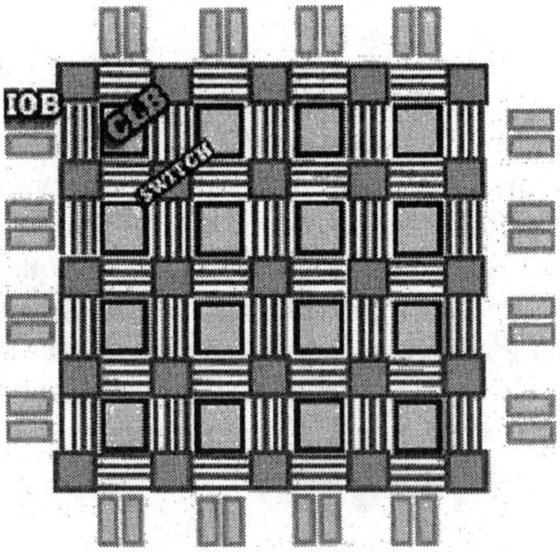

Fig. 7.1. General structure of an island-based FPGA

MFSs design flow has three major tasks: partitioning, placement and routing (see figure 7.3). Frequently two of these tasks are tackled together, because when accomplishing the partitioning, the placement must be considered or vice versa in order to obtain the optimal implementation. In this chapter a methodology, based on evolutionary computation, for the automation of the

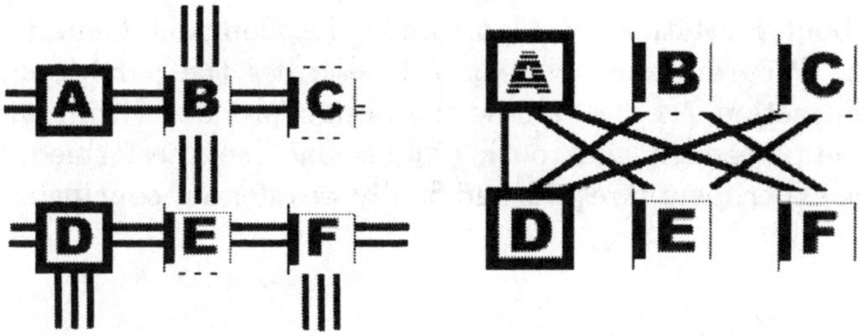

(a) crossbar (b) topologies

Fig. 7.2. Multi-FPGA Mesh

whole design flow is explained. There are two separated steps: First, the partitions of the circuit are obtained. During the first stage of the design flow, we also assign a partition (portion of the circuit) to each FPGA. The second step is devoted to place and route the circuit using the FPGA resources. Two different evolutionary algorithms are used: a hybrid compact genetic algorithm (HcGA) for the partitioning step and the genetic programming (GP) technique for the routing and placement step. The experimental results have been obtained in the basis of a real board made up of 8 FPGA (see later 7.11).

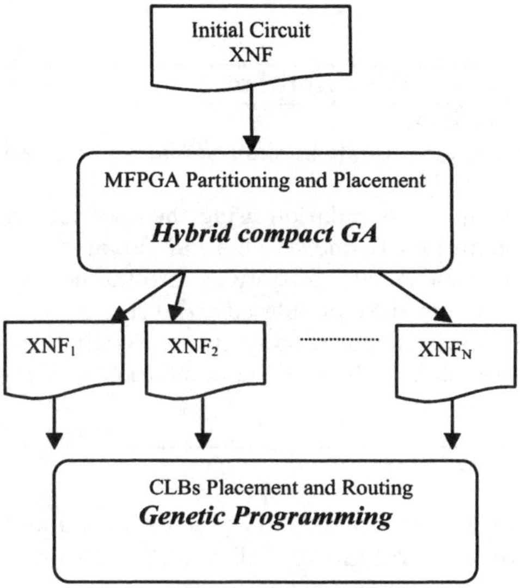

Fig. 7.3. MFS Design Flow

The rest of the chapter is organized as follows: section 7.2 shows an overview about Evolutionary Algorithms, the Compact Genetic Algorithm and Genetic Programming. Section 7.3 describes the partitioning methodology, while section 7.4 shows how the design process within the FPGAs - including the placement and routing steps- has been performed. Section 7.5 contains the experimental results and finally we offer our conclusions in section 7.6.

7.2 Evolutionary Algorithms

Several decades ago, some researchers begun to explore how some ideas taken from nature could be adapted and harnessed for solving well-know difficult problems. Among the concepts borrowed from nature, natural evolution demonstrated from the beginning how simple but also brittle ideas can be helpful for devising new ways of solving difficult problems. Among the techniques that arose under the umbrella of natural evolution, Genetic Algorithms (GAs) [9], Evolutionary Programming [10] and Evolution Strategies [11, 12] have pioneered, matured and demonstrated its usefulness. More recently, John Koza [13] presented Genetic Programming (GP) a new technique that aims at automatically developing computer programs. Koza employed Lisp expressions for evolving programs, and this has favored the use of tree-like data structures in GP, although some researchers have sometimes employed different alternatives. Basically, any EA -including GP- can be described by means of algorithm 7.1.

Algorithm 7.1 Evolutionary algorithm

1. Initialize the population.
2. Evaluate all of the individuals in the population and assign a fitness value to each one.
3. Select individuals in the population using the selection algorithm.
4. Apply genetic operations to the selected individuals.
5. Insert the result of the genetic operations into the new population.
6. If the population is not fully populated go to step 3.
7. If the termination criterion is reached, then present the best individual as the output. Otherwise, replace the existing population with the new population and go to step 3.

We notice from the algorithm that an evaluation process is performed in step 2. Therfore, for evaluating individuals, a fitness function has to be implemented. This function is in charge of computing a fitness value for the individual under evaluation. The fitness value is proportional to the quality of the individual. The selection operation usually takes into account the fitness value of individuals, and select with higher probabilities those with larger fitness values. Finally, we must point out that crossover and mutation are the genetic operations applied to the individuals selected. Crossover operator takes a couple of individuals, that act like parents, and exchange some of their information, thus creating a couple of new descendant individuals, that share information from both parents. On the other hand, the mutation operation, randomly mutate some of the information contained in the individual to which the operation is applied. Depending on the kind of EA employed, different data structures for encoding candidate solutions -individuals- might be employed. Typically, individuals are encoded by means of bit or integer strings when using GAs, while tree structures are employed for GP.

7.2.1 The Compact Genetic Algorithms

In [14] a compact Genetic Algorithm (cGA) has been proposed. It does not manage a population of solutions but only mimics its existence and it simulates the order-one behavior of a simple GA with uniform crossover. The cGAs' authors do not propose it as an alternative algorithm but it can be used to quickly estimate the "difficulty" of a problem. A problem is easy if it can be solved with a cGA exploiting a low selection rate. The more the selection rate must be increased to solve the problem, the more it has to be considered difficult.

The idea on which the cGA is based was primarily inspired by the *random walk* model, proposed to estimate GA convergence on a class of problems in which there is no interaction among the building blocks constituting the solution [15]. Other concepts that inspired the cGA were *Bit-based Simulated Crossover* (BC) [16] and *Population-Based Incremental Learning* (PBIL) [17]. The cGA represents the population by means of a vector of values $p_i \in [0,1], \forall i = 1, \ldots, l$, where l is the number of alleles needed to represent the solutions. Each value p_i measures the proportion of individuals in the simulated population which have a zero (one) in the ith locus of their representation. By treating these values as probabilities, new individuals can be generated and, based on their fitness, the probability vector updated accordingly in order to favour the generation of better individuals.

The initial probabilities values, p_i, are set to 0.5 to represent a randomly generated population in which the value for each allele has equal probability. At each iteration, the CGA generates two individuals on the basis of the current probability vector and compares their fitness. Lets W be the representation of the individual with better fitness, and L the one of the individual whose fitness was worse. The competitor representations are used to update the probability vector at step $k+1$ in the following way:

$$p_i^{k+1} = \begin{cases} p_i^k + 1/n & \text{if } W_i = 1 \wedge L_i = 0 \\ p_i^k - 1/n & \text{if } W_i = 0 \wedge L_i = 1 \\ p_i^k & \text{if } W_i = L_i \end{cases} \tag{7.1}$$

where n is the dimension of the population simulated, and W_i (L_i) is the value of the ith allele of W (L). The cGA ends when the values of the probability vector are all equal to 0 or 1. At this point the vector p itself represents the final solution. Note that the cGA evaluates an individual by considering its whole chromosome. At each iteration, some alleles of solution W might not belong to the optimal solution of the problem, and the correspondent probability values wrongly modified. For example, consider the OneMax problem, in which the related fitness function computes the number of bits set to 1 of a binary string. Lets $a = 10110$ and $b = 01010$ be the two competitors. String a clearly is the individual with better fitness. The first and third element of the probability vector are thus increased by $1/n$, the fourth and fifth elements remain unchanged, while the second element is incorrectly decreased by $1/n$.

Algorithm 7.2 Pseudo-code of the CGA for the TSP.

```
Program TSP_CGA
  begingroup
    Initialize(P,method);
    F_best := INT_MAX;
    count := 0;
    repeat
      S[1] := Generate(P);
      F[1] := Tour_Lenght(S[1]);
      idx_best := 1;
      for k := 2 to s do
        S[k] := Generate(P);
        F[k] := Tour_Lenght(S[k]);
        if (F[k] < F[idx_best]) then idx_best := k;
      end for
      for k := 1 to s do
        if (F[idx_best] < F[k]) then Update(P,S[idx_best],S[i]);
      end for
      if (F[idx_best] < F_best) then
        count := 0;
        F_best := F[idx_best];
        S_best := S[idx_best];
      else
        Update(P,S_best,S[idx_best]);
        count := count + 1;
      end if
    until (Convergence(P) OR count > CONV_LIMIT)
    Output(S_best,F_best);
  end
```

In order to represent a given population of n individuals, the cGA updates the probability vector by a constant value equal to $1/n$. Only $\log_2 n$ bits are thus needed to store the finite set of values for each p_i. The CGA therefore requires $\log_2 n * l$ bits with respect to the $n * l$ bits needed by a classic GA. Larger population dimension can be exploited without significantly increasing memory requirements, but only slowing CGA convergence. This peculiarity makes the use of CGAs very attractive to solve problems for which the huge memory requirements of GAs is a constraint.

To solve problems higher than order-one GAs with both higher selection rates and larger population sizes have to be exploited [18]. The cGA selection pressure can be increased by modifying the algorithm in the following way: **(1)** generate at each iteration s individuals from the probability vector instead of two; **(2)** choose among the s individuals the one with best fitness and select as W its representation; **(3)** compare W with the other $s-1$ representations and update the probability vector accordingly. The other parts of the algorithm remain unchanged. Such an increase on the selection pressure helps the cGA to converge to better solutions since it increases the survival probability of higher order building blocks [14]. Algorithm 7.2 shows a pseudocode of the cGA for the TSP problem.

7.2.2 Genetic Programming

One of the difference between GP and other EAs is that fitness values are to be computed by evaluating computer programs. If we consider that individuals -programs- are encoded by means of tree like structures (see figure 7.4), each program is made up of internal nodes -functions- and terminals -the leaves of the tree. Which functions and terminals are of interest for the problem that is to be solved is decided by the researcher, and usually varies largely from a problem to another. For instance, if we employ GP for solving a symbolic regression problem, we may choose arithmetic functions for the function set, while if we apply GP for programming a robot, some primitives that allows to move the robot along several directions could make up the function set. The terminal set are usually made up of the constant values and parameters employed by the functions included in the terminal set. Therefore, the first concern for GP practitioners is to appropriately define the function and terminal sets. This means that even when the solution for the problem to be addressed is not known, one must be sure that the solution can be found using the functions and terminals selected.

Genetic operators applied in GP are similar to those employed with any other Evolutionary Algorithm. One of the main differences is due to the kind of data structures employed. When crossover is applied to a couple of individuals, two new descendants are obtained by exchanging some randomly chosen subtrees from each of the parents (see figure 7.5). On the other hand, mutation operator generates a new individual by substituting a randomly chosen subtree from the parent, by a new one that is also randomly generated (see figure 7.6). Although other possibilities are available, the previously described ones are the simplest and most widely employed versions of the genetic operators.

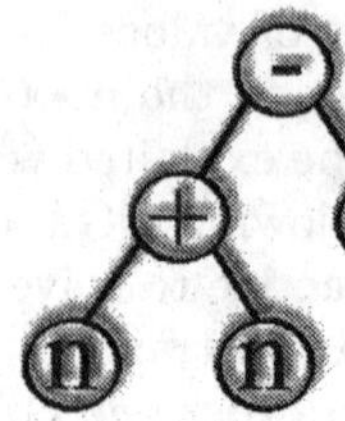

Fig. 7.4. Individuals are encoded by means of trees in Genetic Programming.

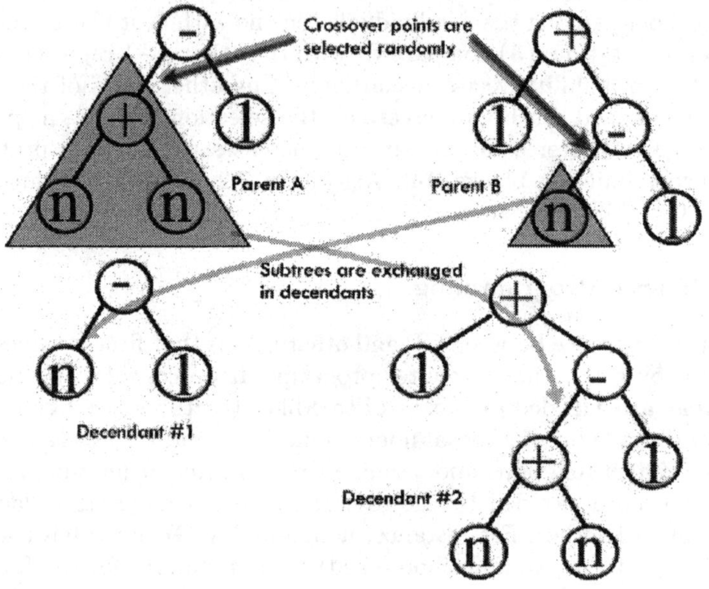

Fig. 7.5. Crossover operation.

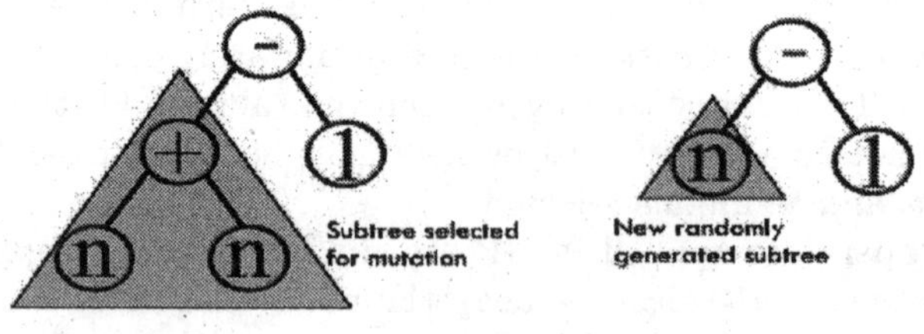

Fig. 7.6. Mutation operation.

Once all of the above components are integrated within the GP algorithm -that is basically the same described in algorithm 7.1-, it can be applied to any optimization problem. In section 7.4 we show how GP has been applied for solving the problem of Placement and routing circuits on FPGAs. A wider description of Genetic Programming can be found in [19].

7.3 MFS partitioning and FPGA assignment

In this section we present the first stage of the design flow. We describe different techniques and algorithms presented in several papers. Most of the previous approximations do not preserve the structure of the circuit or use a difficult encoding. For example Laszewski and M1/4hlenbein implemented a parallel GA which solves the graph partitioning problem with an easy encoding, but the solutions do not preserve the structure of the circuit, and that is a key issue if we want to minimize the delays of the partitioned circuitn [20]. Alpert uses a GA for improving another partitioning algorithm with good results for bi-partitions [21] . An exception, concerning the structure, is the approximation made by Hulin [22]. The approximation used here solves these problems. It is adaptable and can be modified for using in other graph partitioning problems with few changes, it is parallelizable (the method is intrinsically parallel, because it uses a genetic algorithm as a tool for optimization), and in addition, the evaluation of the fitness function can be parallelized very easily. The algorithm also preserves the structure of the circuit and it detects those parts of the graph which are independent.

7.3.1 Methodology

partitioning deals with the problem of dividing a given circuit into several parts, called partitions, in order to be implemented on a MFS. The partitions are obtained and each partition is assigned to a different FPGA within the board. We use a 8-FPGA Mesh topology board, so we must bear in mind several constraints related to the board. Some, and usually most important, of these constraints are the number of available I/O pins on each FPGA and logic capacity. FPGA devices have a much reduced number of pins when compared with their logic capacity. In addition we must connect parts of the circuit that are placed on non-adjacent FPGAs, and for this task we have to use some of the available pins. Partitioning appears in a lot of design automation design problems, and most of the research related to MFS partitioning were adapted from other VLSI areas [23]. For this specific board we have developed a new methodology. We apply the graph theory to describe a given circuit, and then a compact genetic algorithm (cGA) with a local search improvement is applied with a problem-specific encoding. This algorithm not only preserves the original structure of the circuit but also evaluates the I/O-pins consumption due to direct and indirect connections between FPGAs. The MFS placement or FPGA assignment is done by means of a fuzzy technique. We have used the partitioning93 benchmarks [24], described in the Xilinx Netlist Format (XNF), a netlist description language [25].

7.3.2 Circuit Description

Some authors use hyper-graphs as the way of representing a circuit, but there are also some approximations, which use graphs [26]. We have thus, employed an undirected graph representation to describe the circuit. This representation permits an efficient encoding of the compact genetic algorithm and a direct encoding of the solutions using this code.

Hidalgo et al. [27] describe a method that uses the edges of a graph to represent k-way partitioning solutions. They transform the netlist circuit description into a graph, and then operate with its spanning tree. A spanning tree of a graph is a tree, which has been obtained selecting edges from this graph. One of the properties of a spanning tree is that if n edges are suppressed, $n-1$ isolated trees are obtained. As we are treating a k-way partitioning problem, $k-1$ edges of the spanning tree are selected and eliminated in order to obtain k partitions of the original circuit. The partitions are represented by the deleted edges and a hybrid compact genetic algorithm (HcGA) works under this representation to obtain the best partitioning accordingly to the board constraints previously explained. Based on the previous statement, a specific algorithm to address the partitioning and placement problems in MFS systems can be used. The algorithm, which is also adaptable to different boards and devices, preserves the main structure of the circuit and, by means of a fuzzy technique, evaluates the IO pins consumption due to not only direct, but also indirect connections between FPGAs within the MFS (an 8-FPGA board).

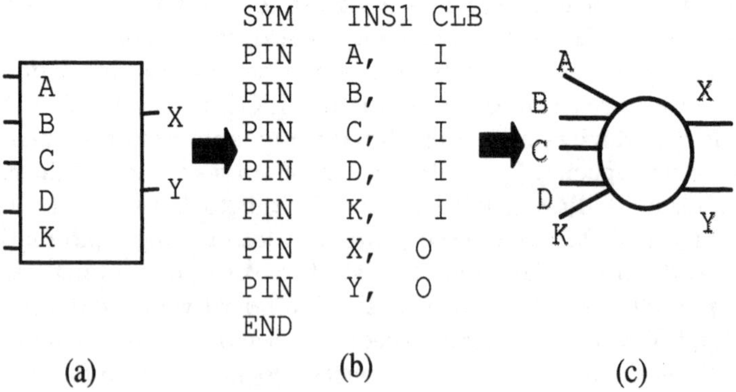

Fig. 7.7. An example of a CLB described in (a)block, (b)XNF,and (c)graph formats.

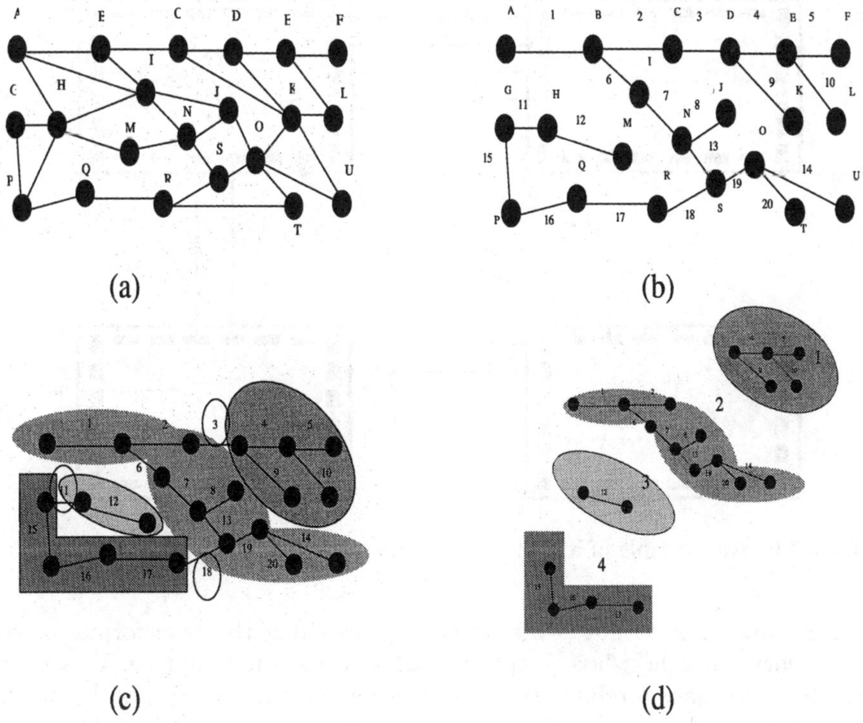

Fig. 7.8. An example of the partitioning process for 4 FPGAs.

The main objective is to solve the circuit partitioning problem and to obtain a set of portions or partitions of the original circuit suitable for the implementation over a single FPGA. The partitioning process is targeted to a device board which has their devices connected in a 4-way mesh topology [2]. So, the method works as follows. First a graph representing the circuit netlist description is obtained. Fig. 7.7 shows the equivalence between an XNF netlist description of a Configurable Logic Block and a graph. After that a spanning tree of that graph is randomly selected, from this tree we select $k-1$ edges and we eliminate them in order to obtain a $k-way$ partition. The partitions are represented by the deleted edges. In Fig. 7.8 we can see an example of the partitioning process. Starting from the circuit graph (a), we get its spanning tree (b) using the Kruskal algorithm [26]. From it, we select the necessary edges and finally we obtain the partitions (c). The figure represents an example for four FPGA devices, so we select only 3 edges of the tree. Once the partitions have been obtained the graph representation can be transformed into a XNF file for each partition and then these files, with the necessary additional information, can be implemented on each FPGA (see Fig. 7.9).

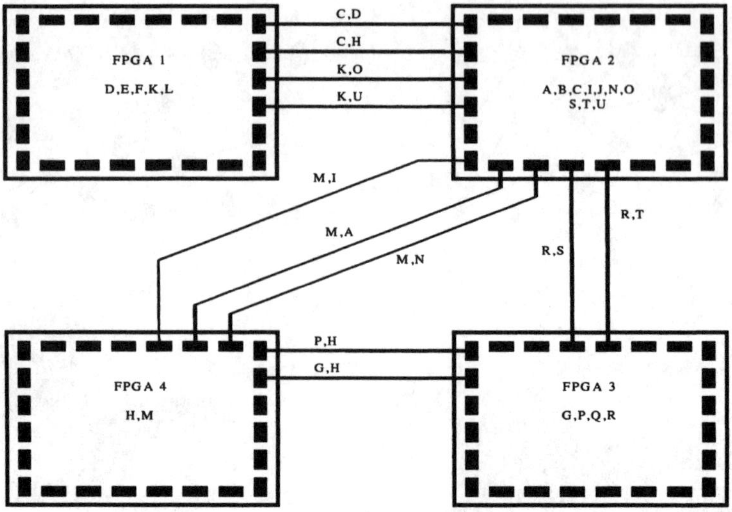

Fig. 7.9. An example of a post-partitioning implementation using 4 FPGAs.

It is important to note that when accomplishing the transformation we should work with the whole graph instead with its spanning tree. This is because the information related to connections is included in the graph and the spanning tree only works with some of them. It is necessary to determine the optimum distribution of the CLBs on the different available FPGAs. An optimum distribution has a minimal cost and guarantees the internal routability of each FPGA.

7.3.3 Genetic Representation

The evaluation process tell us the goodness of the solutions by means of a fitness function. The main task of the HcGA is to solve the partitioning while attending some board requirements related to IO pins and logic blocks (called CLBs on Xilinx's devices). The fitness function guides the search of the algorithm, so it must minimize the number of cutting edges (that is the connections between FPGAS of the MFS) and in addition, it must distribute the blocks uniformly among the FPGAs. So we have a multi-objective genetic algorithm problem. This problem is well known and a number of non-genetic and genetic algorithms have been implemented for its resolution [28] [29]. One of the techniques commonly used is the use of added functions which include weighted sum methods, where the user assigns a weight to each objective and the total fitness is the sum of all weighted fitness values. Nowadays a lot of multi-objective techniques are available for the designer to adapt those partitioning problems.

In order to design a cGA for Multi-FPGA Partitioning we adopted the *edge* representation previously commented and we consider the frequencies of the edges occurring in the simulated population. A vector V of dimension equal to the number of nodes minus one was used to store these frequencies. Each element v_i of V represents the proportion of individuals whose partition use the edge e_i. The vector elements v_i were initialized to 0.5 to represent a randomly generated population in which each edge has equal probability to belong to a solution. In Algorithm 7.3 the pseudo code of a cGA to solve Multi-FPGA partitioning is shown.

Algorithm 7.3 Pseudo-code of the cGA for Multi-FPGA Partitioning.

```
Program Multi-FPGA-cGA
  begin
    Initialize(V);
    F_best := INT_MAX;
    count := 0;
    repeat
       S[1] := Generate(V);
       F[1] := Partition(S[1]);
       idx_best := 1;
       for k := 2 to s do
          S[k] := Generate(V);
          F[k] := Partition(S[k]);
          if (F[k] < F[idx_best]) then idx_best := k;
       end for
       for k := 1 to s do
          if (F[idx_best] < F[k]) then Update(V,S[idx_best],S[i]);
       end for
       if (F[idx_best] < F_best) then
          count := 0;
          F_best := F[idx_best];
          S_best := S[idx_best];
       else
          Update(V,S_best,S[idx_best]);
          count := count + 1;
       end if
    until (Convergence(V) OR count > CONV_LIMIT)
    Output(S_best,F_best);
  end
```

After the initialization phase an individual is generated and its fitness value is computed. Then, according to the selection pressure adopted $s-1$ individuals are generated, evaluated and the best individual is carried out. The last is used to update the probability vector V according to Equation 7.1. Moreover, the best individual generated in the current iteration (S[idx_best]) is compared with the best individual found until now (S_best) and V is updated accordingly. The cGA proposed in [14] ends when the values of the probability vector are all equal to 0 or 1. Since in our tests such a condition was rarely achieved we introduced a supplementary end condition which limits the maximum number of generations occurring without an improvement of the best solution achieved (see algorithm 7.3). Reached such a limit the execution is terminated and the best individual found is returned as final solution.

The cGA (and also the HcGA) uses the encoding presented in section 7.3.2 which directly represents solutions to the partitioning problem. As we have said, the code is based on the edges of a spanning tree. We have seen above how the partition is obtained by the elimination of some edges. A number is assigned to every edge of the tree. Consequently, for a k-way partitioning problem a chromosome will have k-1 genes, and the value of these genes can be any of the order values of the edges. For example, chromosome (3 14 26 32 56 74 89) for a 8-way partitioning, represents a solution obtained after the suppression of edge numbers 3, 14, 26, 32, 56, 74, and 89 from a spanning tree. So the alphabet of the algorithm is: $\Omega = \{0, 1 \ldots, n-1\}$ where n is the number of vertexes of the target graph (circuit), because the spanning tree has $n-1$ edges.

7.3.4 Hybrid Compact Genetic Algorithm

A Hybrid cGA (HcGA) uses non-evolutionary algorithms for local search, that is, to improve good solutions found by the cGA. When designing a cGA for MFS partitioning, a vector (V), with the same dimension as the number of nodes minus one, stores the frequencies of the edges occurring in the simulated population. Each element vi of V represents the proportion of individuals whose partition use the edge ei. Following the original cGA, the vector elements vi were initialised to 0.5 to represent a randomly generated population in which each edge has equal probability to belong to a solution [14]. Sometimes it is necessary to increase the selection pressure rate Ps, (the number of individuals generated on each iteration) to reach to good results with a Compact Genetic Algorithm. A value for Ps near to 4 has shown to be a good value for MFS partitioning. It is not to be recommended a large increasing of this value, because the computation time will grow drastically. Additionally, for some problems we need a complement to cGA in order to solve them properly. We can combine heuristics techniques with local search algorithms to obtain this additional tool called hybrid algorithms. We have implemented a cGA with local search.

In [30] a compact genetic algorithm for MFSs partitioning was presented, and in [31] a Hybrid cGA was explained. Authors combine a cGA with the Lin-Kernighan (LK) local search algorithm, to solve Traveling Salesman Problems (see Algorithm 7.2). The cGA part explores the most interesting areas of the search space and LK task is the fine-tuning of those solutions obtained by cGA. Following this structure, but changing the local search method, we can implement a hybrid cGA for MFS partitioning. Ideally, a local search algorithm must try to perform the search process as exhaustively as possible. Unfortunately, in our problem this also implies an unacceptable amount of computation. Therefore, we have employed a local search heuristic each certain number (n) of iterations and we need to study the value of n to keep the algorithm search in good working order. After empirically studying the local search frequency, we have obtained that n must be assigned a value between 20 and 60, with an optimal value (that depends on the circuit benchmark) near to 50. So for our experiments we fixed the local search frequency n to 50 iterations, i.e. we develop a local search process every 50 iterations of the cGA.

Now it is necessary to define a new concept, neighbouring. We have mentioned that a chromosome has $k-1$ genes for a k-way partitioning, and the value of these genes are the edges that are removed from the spanning tree representing the circuit when looking for a solution.

Definition.

solution A is a neighbour solution of B (and B is a neighbour solution of A) if the difference between their chromosomes is just one gene.

Our local search heuristic explores only one neighbour solution for each gene, that is $k-1$ neighbouring solutions of the best solution every n iterations. The local search process works as Algortihm 7.4 explain [32].

Although only a very small part of the solution neighbourhood space is explored, the performance of the algorithm improves significantly (in terms of quality of solutions) without degrading drastically its total computation time. In order to clarify the explanation about the proposed local search method we can see an example. Let us suppose a graph with 12 nodes and its spanning tree, for a 5-way partitioning problem (i.e. we want to divide the circuit into five parts). As we have explained, we will use individuals with 4 genes. Let us also suppose a local search frequency (n) of 50 and that after 50 iterations we have reached to a best solution represented by:

$$BS = (3, 4, 6, 7) \tag{7.2}$$

The circuit graph has 12 nodes, so its spanning tree is formed by 11 edges. The whole set of possible edges to obtain a partitioning solution is called E:

$$E = \{0, 1, 2, 3, 4, 5, 6, 7, 8, 9, 10\} \tag{7.3}$$

In order to generate TS1 we need to know the available edges ALS for random selection, as we have said, we eliminate the edges within BS from E to obtain ALS:

$$ALS = \{0, 1, 2, 5, 8, 9, 10\} \tag{7.4}$$

Algorithm 7.4 Local search algorithm for MFS partitioning HcGA.

1. Every n iterations, we obtain the best solution up to that time (*BS*).To obtain *BS*:
   a) first we explore the compact GA probability vector and select the k-1 most used genes (edges) to form *MBS* (vector best individual).
   b) The best individual generated up to now (*GBS*) (similar to elitism) is also stored.
   c) The best individual between *MBS* and *GBS* (i.e. which of them has the best fitness value) will be *BS*.
2. the first random neighbour solution (*TS1*) to *BS* is generated substituting the first gene (edge) of the chromosome by a random one, not present in *BS*.
3. Calculate the fitness value of *BS* ($FVBS$) and the fitness value of *TS1* ($FVTS1$)
4. Compare If *FVTS1* is better than *FVBS*, if so *TS1* is dropped to *BS* and the initial *B*S is eliminated, otherwise *TS1* is eliminated
5. Repeat the same process using the *new BS* and with the second gene, to generate *TS2*
6. If the fitness value of *TS2* ($FVTS2$) is better than the present $FVBS$ then *TS2* will be our *new BS* or, if *FVTS2* is worst than $FVBS$, there will be no change in *BS*.
7. Repeat last step for the rest of the genes untill the end of the chromosome (that is, k-1 times for a k-way partitioning).

Now we randomly select an edge (suppose 0) to build TS1substituting it by the first gene in BS:

$$TS1 = (0, 4, 6, 7) \tag{7.5}$$

The third step is the evaluation of $TS1$ (suppose $FVTS1 = 12$) and comparing (suppose a minimization problem) with $FVBS$ (suppose $FVBS = 25$). As FVTS1 is better than FVBS, TS1 will be our new BS and the original BS is eliminated. Those changes also affect to ALS because our new ALS is:

$$ALS = \{1, 2, 3, 5, 8, 9, 10\} \tag{7.6}$$

Table 7.1 represents the rest of the local search process for this example.

7.4 Placement and Routing on FPGAs

Once the first step has been carried out, we have several partitions. Each partition - that is in charge of a small circuit - have to be implemented independently in a different FPGA. Finally, all the FPGAs will be connected together, thus obtaining the global circuit. Even when much research has been done on the automatic generation of digital and analogue circuits, we will review now some proposals that are related with the idea of applying evolutionary algorithms to the problem we are addressing, and with the way circuits are encoded.

Table 7.1. Local Search example

i	ALS	BS	FV	Random gene	TS	FV	New Bs
1	0,1,2,5,8,9,10	3,4,6,7	25	0	0,4,6,7	12	0,4,6,7
2	1,2,3,5,8,9,10	0,4,6,7	12	1	0,1,6,7	37	0,1,6,7
3	1,2,3,5,8,9,10	0,4,6,7	12	9	0,4,9,7	10	0,4,9,7
4	1,2,3,5,6,8,10	0,4,9,7	10	8	0,4,8,9	11	0,4,9,7
	Pre-Local Search	Best Solution:		3,4,6,7			
	Post-Local Search	Best Solution:		0,4,9,7			

A given circuit, with wires, gates and connections, can be considered as a graph. Several papers have dealt with the problem of encoding graphs, i.e. circuits, when working with GA and GP [33]. Sometimes new techniques have been developed to do so. For instance, Cartesian Genetic Programming [34] is a variation of GP which was developed for representing graphs, and shows some similarities to other graph based forms of genetic programming. Miller et al's aim is to find complete circuits capable of implementing a given boolean function. Nevertheless, we are more interested in physical layout. Our optimisation problem begins with a given circuit description, and the goal is to find out how to place components and wires in FPGAs. Meanwhile we have also developed a new methodology for representing circuits by means of GP with individuals represented as trees.

Other researchers have also applied Evolutionary Algorithm for evolving analogue circuits [33]. Even Koza have employed Genetic Programming for designing and discovering analogue circuits [35], which have eventually been patented. Thompson's research scope is the physical design and implementation of circuits in FPGAs [36]. However, all of them work with analogue circuits, while we are addressing digital ones. Another difference is the kind of evolutionary algorithm employed for solving each problem. Thompson uses GAs while we are using GP (Koza uses GP but not for solving the kind of problem we address here).

There are also other researchers that have addressed problems employing reconfigurable hardware and Genetic Programming. For instance, in [37] authors describe how trees can be implemented and evaluated on FPGAs. But our aim is not to implement a Genetic Programming tool on an FPGA but using GP for physically placing and routing circuits. Therefore, in this second step, we take each of the partitions as the input of the problem, and the goal is to place components and establish connections among them in a different FPGA. Our proposal now is to use Genetic Programming (GP) for solving this task. The main reason behind this choice is the similarity between data structures that GP uses -trees- and the way of describing circuits -graphs. A tree is more convenient than a fix-sized string for describing graphs of any length. In the following sections we describe how graphs are encoded by means of trees.

7.4.1 Circuits encoding using trees

As described in section 7.3, the output for the partitioning algorithm is a set of partitions, and a description of the way they must be connected. Each of the partition includes a circuit that must be implemented in a separate FPGA. Therefore, the main goal for this step is to implement a partition (circuit) into an FPGA. Each of the circuit component has to be implemented into a CLB, and after that previous step, all the CLBs have to be connected according to the circuit's topology. Given that we use tree-based GP in this stage of the methodology, we need a mapping between a graph -circuit- and a tree. Circuits have to be encoded as trees, and any of the trees that GP will generate, should also have an equivalent circuit; the fitness function will later decide if the circuit is correct or not, and its resemblance degree with the correct circuit.

Considering that any of the components of a circuit is simple enough to be implemented employing a CLB from the FPGA, we might describe a circuit employing black boxes, such as is depicted by means of an example in figure 7.10. This means that we only have to connect CLBs from the FPGA according to the interconnection model that a given circuit implements, and then we can configure each of the CLB with the function that each component performs in the circuit. We want to perform this task by using GP. This means that circuits must be described by means of trees -individuals in GP. To do it, we can firstly label each component from the circuit with a number, and then assign components' labels to the ends of wires connected to them (see figure 7.10).

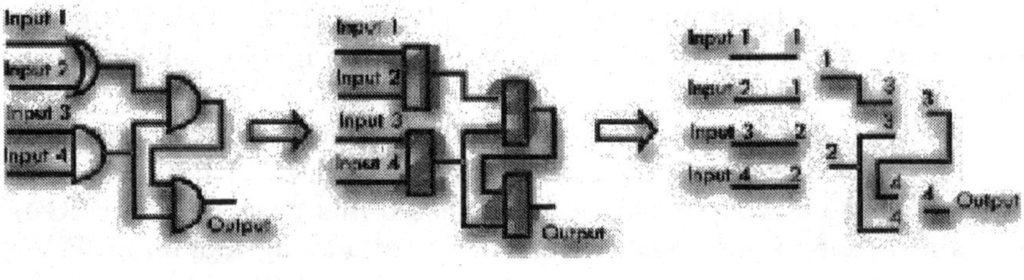

Fig. 7.10. Representing a circuit with black boxes.

We may now describe all the wires by means of a tree by connecting each of the wires as a branch of the tree and keeping them all together in the same tree. By labeling both extremes of branches, we will have all the information required to reconstructing the circuits. Any given tree, randomly generated, will always correspond to a particular graph, regardless of the usefulness of the associated circuit (see figure 7.8). In this proposal, each node from the tree is representing a connection, and each branch is representing a wire. The next stage is to encode the path of wires into an FPGA. Each branch of the tree will encode a wire from the circuit: internal nodes specify switch connections that are traversed by the wire, while the first and last nodes of the branch are employed to connect the wire to an adjacent CLB -by specifying which of the CLB is employed and to which pin is the wire connected.

Each of the branches will include as many internal nodes as required for describing all of the switch connections required for the wire (see figure 7.8). Sometimes, branches will not include any internal nodes. This may happen when an input/output connection is directly attached to any of the CLB from the surrounding area of the FPGA. Only two nodes are required in the branch: the first one specify which IOB is employed, while the second one select the CLB to which it is connected and the wire employed.

Each internal node requires some extra information: if the node corresponds to a CLB we need to know information about the position of the CLB in the FPGA, the number of pin to which one of the ends of the wire is connected, and which of the wires of the wire block we are using; if the node represents a switch connection, we need information about that connection (figures 7.11 and 7.12 graphically depicts how a tree describes a circuit, and the way each branch maps a connection).

It may well happen that when placing a wire into an FPGA, some of the required connections specified in the branch can not be made, because, for instance, a switch block connection has been previously used for routing another wire segment. In this case the circuit is not valid, in the sense that not all the connections can be placed into a physical circuit, and the function in charge of analyzing the tree will apply a high penalty to that individual from the population.

In order for the whole circuit to be represented by means of a tree, we will use a binary tree, whose left most branch will correspond to one of its connections, and the left branch will consist of another subtree constructed recursively in the same way (left-branch is a connection and right-branch a subtree). The last and deepest right branch will be the last circuit connection. Given that all internal nodes are binary ones we can use only a kind of function with two descendants. In the following subsection we describe the GP sets required.

7.4.2 GP sets

When solving a problem by means of GP one of the first things to do once the problem has been analyzed is to build both the function and terminal sets. The function set for our problem contains only one element: F={SW}, Similarly, the terminal set contains only one element T={CLB}. But SW and CLB may be interpreted differently depending on the position of the node

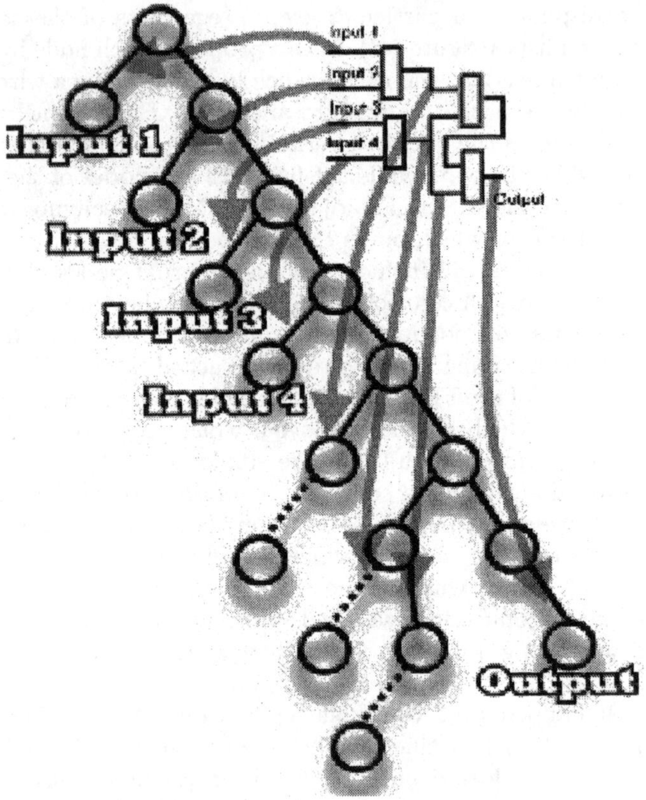

Fig. 7.11. Making connections in the FPGA according to nodes

within a tree. Sometimes a terminal node corresponds to an IOB connection, while sometimes it corresponds to a CLB connection in the FPGA (see figure 7.8. Similarly, an internal node - SW node- sometimes corresponds to a CLB connection (the first node in the branch), while others affects switch connections in the FPGA (internal node in a branch, see figure 7.9). Each of the nodes in the tree will thus contain different information:

- If we are dealing with a terminal node, it will include information about the position of CLBs, the number of pins selected, the number of wires to which it is connected, and the direction we are taking when placing the wire.
- If we are instead in a function node, it will have information about the direction we are taking. This information enables us to establish the switch connection, or in the case of the first node of the branch, the number of the pin where the connection ends.

We can notice in figure 7.8, that wires with IOBs at one of their ends are shorter -only needs a couple of nodes- than those that have CLBs at both

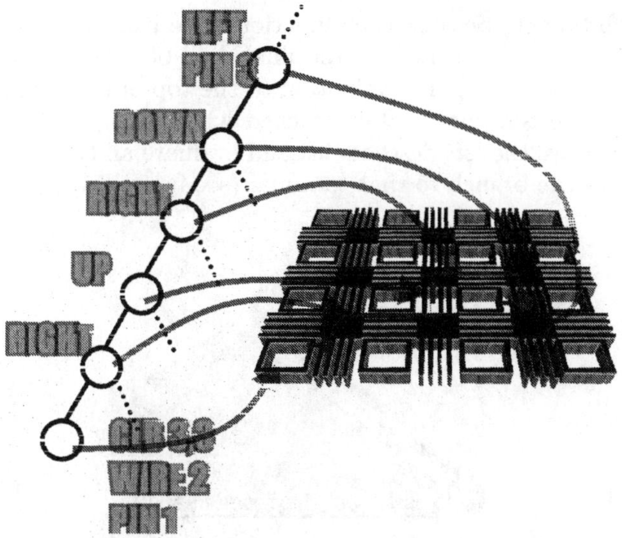

Fig. 7.12. Encoding circuits by means of binary trees. Each branch of the tree describes a connection from the circuit. Dotted lines indicates a number of internal nodes in the branch

ends -they require internal nodes for expressing switch connections-. Wires expressed in the latest position of trees have less space to grow, and so we decided to place IOB wires in that position, thus leaving the first parts of the trees for long wires joining CLBs.

7.4.3 Evaluating Individuals

In order for GP to work, individuals from the population have to be evaluated and reproduced employing the GP algorithm. For evaluating an individual we must convert the genotype (tree structure) to the phenotype (circuit in the FPGA), and then compare it to the circuit provided by the partitioning algorithm. We developed an FPGA simulator for this task. This software allows us to simulate any circuit and checks its resemblance to other circuit. Therefore, this software tool is in charge of taking an individual from the population and evaluating every branch from the tree, in a sequential way, establishing the connections that each branch specifies. Circuits are thus mapped by visiting each of the useful nodes of the trees and making connections on the virtual FPGA, thus obtaining phenotype. Each time a connection is made, the position into the FPGA must be brought up to date, in order to be capable of making new connections when evaluating the remaining nodes. If we evaluate each branch, beginning with the terminal node, thus establishing the first end of the wire, we could continue evaluating nodes of the branch from the bottom to the top. Nevertheless, we must be aware that there are several terminals related to each branch, because each function node has two different descendants. We must decide which of the terminals will be taken as the beginning of the wire, and then drive the evaluation to the top of the branch. We have decided to use the terminal that is reached when going down through the branch using always the left descendant, and evaluate all the nodes traversed from the root of the branch to that terminal (see figure 7.13).

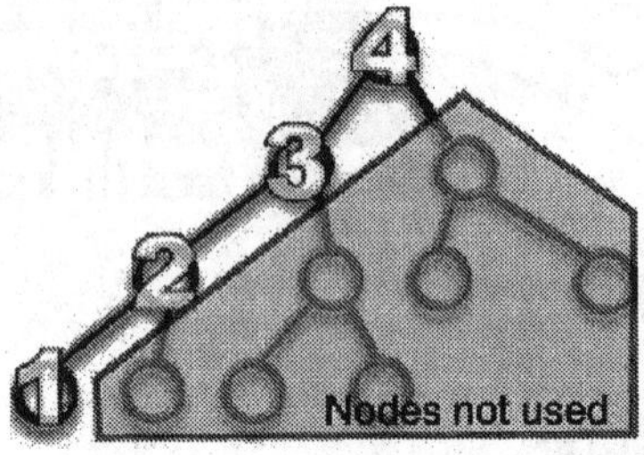

Fig. 7.13. Evaluating a branch of the tree-corresponding to a connection of the circuit. Evaluation order is specified with numbers labelling nodes.

In one sense there is a waste of resources when having so many unused nodes. Nevertheless they represent new possibilities that can show up after a crossover operation (in nature, there always exist recessive genes, which from time to time appear in descendants). These nodes are hidden, in the sense that they do not take part in the construction of the circuit and may appear in new individuals after some generations. If they are useful in solving the problem, they will remain in descendants in the form of nodes that express connections. The fitness function is computed as the difference between the circuit provided and the circuit described by the individual.

7.5 Experimental Results

7.5.1 partitioning and Placement onto the FPGAs

The algorithm has been implemented in C and run on a Pentium 3, 866 MHz with Linux Red Hat 7.3. We have used the MCNC partitioning benchmarks in XNF format. We have supposed that each block of the circuits uses one CLB. We use the Xilinx's 4010 FPGA. 7.2 contains the experimental results. It has five columns which express: the name of the test circuit (Circuit), its number of CLBs (CLB), the number of connections between CLBs (Edges), the number of CLBs used on each FPGA (Distribution) and the CPU time in seconds necessary to obtain a solution for 100 generations of a GA with a population of 501 individuals (T(sec)). There are some unbalanced distributions, because we need to use some resources to pass the nets from one device to another. In addition our fitness function has been developed to achieve two objectives, so that the GA works. To cap it all, the algorithm succeeds in solving the partitioning problem with board constraints.

Fig. 7.14 shows a picture of the board. This card consists of 8 FPGAs of the 4010 family from Xilinx [38] although, these can be replaced by other devices of greater capacity and benefits, just adapting the connections. The FPGAs are connected according to a mesh topology, in other words, they directly connect their next neighbours. The figure shows, in addition to the FPGAs, the electrical power supply and lines for programming them (DIN, DONE, CCLK, INIT, PROGRAM), which allows the configuration by means of an *XChequer* cable from Xilinx. The cable transmits the configuration data to all FPGAs within the board, the transmission frequency is 921 kHz. The speed depends on the used computer, in our case with a PC, a Baud Rate of 115200 can be reached. The power supply used is an ATX computer source. This allows us to have the voltages necessary to feed not only the FPGAS, but also the programming cables such as the *XChequer*. The MFS board also incorporates some jumper pins, for programming and isolation of a group of FPGA within the board. There are also six connectors for expansion of the board using other similar card.

Fig. 7.14. Multi-FPGA board designed for testing the methodology

Table 7.2. Experimental Results for Partitioning and Placement for the 8 -Xilinx 4010 Board

Circuit	CLB	Edges	Distribution	T(sec)
S208	127	200	23,9,16,15,16,18,16,14,	7.83
S298	158	306	24,14,23,27,18,19,21,12	11.91
S400	217	412	15,10,25,34,30,44,30,29	15.71
S444	234	442	17,37,41,15,33,27,29,35	20.63
S510	251	455	28,39,44,23,33,29,30,25	17.64
S832	336	808	32,38,33,38,47,25,65,58	166.94
S820	338	796	41,45,49,49,21,33,41,59	43.95
S953	494	882	45,87,58,63,68,73,67,33	34.83
S838	495	800	30,34,73,25,63,70,34,164	173.71
S1238	574	1127	78,65,70,69,80,98,61,53	483,16
C3540	1778	2115	114,116,119,121,118,128,169,153	1634.00

7.5.2 Inter-FPGA Placement and Routing

Several experiments with different sizes and complexities have been performed for testing the placement and routing process . Fig. 7.15 graphically depicts one of the circuits employed in the series of test of increasing complexity that has been used for validating the methodology (a larger set of experiments and results can be found in [39]). The main parameters employed were the following: Number of generations = 500, Population size: 200, Maximum depth: 30, Steady State Tournament size: 10. Crossover probability=98%, Mutation probability=2%, Creation type: Ramp Half/Half, and elitism.

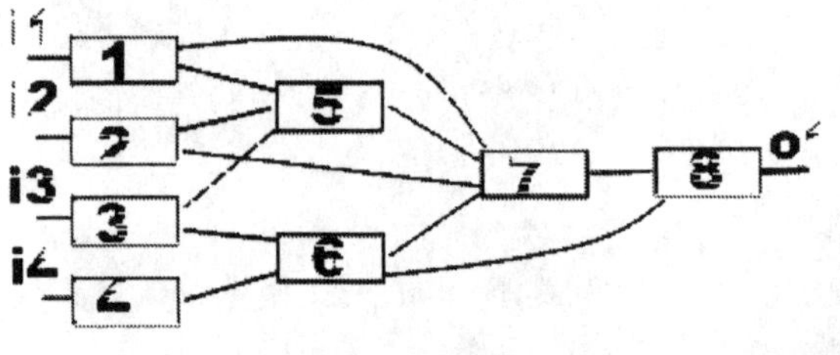

Fig. 7.15. One of the circuits employed for testing the methodology

Fig. 7.16 shows some of the solutions that were obtained with GP- for the circuit described above. A very important fact is that each of the solutions that GP found possesses different features, such as area of the FPGA used, position of the input/output terminals. This means that the methodology could easily be adapted for managing typical constraints in FPGA placement and routing. More solutions found for this and other circuits are described in [39] and [40]. The time required for finding the solution was of some minutes in a 2Ghz Pentium processor. So, the methodology can be successfully employed for routing circuits of larger complexity.

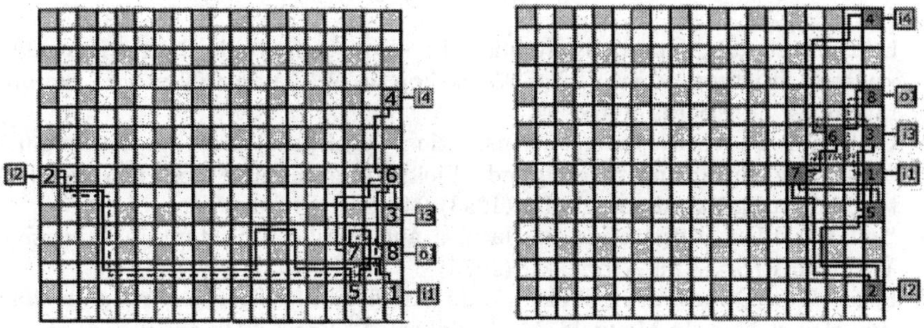

Fig. 7.16. Different solutions obtained by means of GP

7.6 Summary

In this chapter a methodology for circuit design using Multi-FPGA Systems has been presented. We have used evolutionary computation for all the steps of the process. Firstly, an Hybrid compact genetic algorithm was applied on achieving partitioning and placement for inter-FPGA systems and, for the Intra-FPGA tasks Genetic programming was used. This method can be applied for different boards and solves the whole design flow process.

7.7 Acknowledgments

Part of this research has been possible thanks to Ministerio de Ciencia y Tecnologa, research projects number TIC2002-04498-C05-01 and TIC 2002/750.

References

1. S.Trimberger: A reprogrammable gate array and applications. Proceedings of the IEEE **81** (1993) 1030–1040
2. Hauck, S.: Multi-FPGA systems. PhD thesis, University of Washington (1994)
3. Macketanz, R., Karl, W.: Jvx - a rapid prototyping system based on java and fpgas. In Springer-Verlag, ed.: Field Programmable Logic: From FPGAs to Computing Paradigm, Berlin (1998) 99–108
4. Hauck, S.: The roles of fpgas in re-programmable systems. Proceedings of the IEEE **86** (1998) 615–638

5. Baxter, M.: Icarus: A dynamically reconfigurable computer architecture. In: IEEE Symposium on FPGAs for Custom Computing Machines. (1999) 278–279
6. Heywood, M., Zincir-Heywood, A.: Register based genetic programming on fpga computing platforms. In Spinger-Verkag, ed.: Proceedings of EuroGP 2000. (2000) 44–59
7. HArtenstein, R., Kress, R., Reinig, H.: A reconfigurable data-driven alu for xputers. In Press, I., ed.: IEEE Workshop on FPGAs for CUstom Computing Machines. (1994) 139–146
8. Callahan, T., Wawrzynek, J.: Instuction- level parallelism fot reconfigurable computing. In Springer-Verlag, ed.: Field Programmable Logic: From FPGAs to Computing Paradigm, Berlin (1998) 248–257
9. Holland, J.H.: Adpatation in Natural and Artificial Systems. University of Michigan Press, Ann Arbor, MI (1975)
10. Fogel, L.J., Owens, A.J., Walsh, M.J.: Artificial intelligence through a simulation of evolution. In Maxfield, M., Callahan, A., Fogel, L.J., eds.: Biophysics and Cybernetic Systems: Proc. of the 2nd Cybernetic Sciences Symposium, Washington, D.C., Spartan Books (1965) 131–155
11. Rechenberg, I.: Evolutionsstrategie: Optimierung technischer Systeme nach Prinzipien der biologischen Evolution. frommann-holzbog, Stuttgart (1973) German.
12. Schwefel, H.P.: Evolutionsstrategie und numerische Optimierung. PhD thesis, Technische Universitat Berlin, Berlin (1975)
13. Koza, J.R.: Genetic Programming: On the Programming of Computers by Means of Natural Selection. MIT Press, Cambridge, MA, USA (1992)
14. G. Harik, F.L., Goldberg, D.: The compact genetic algorithm. Technical Report 97006, University of Illinois at Urbana-Champaign, Urbana, IL (1997)
15. Harik, G.R., Lobo, F.G., Goldberg, D.E.: The compact genetic algorithm. IEEE-EC **3** (1999) 287
16. Syswerda, G.: Simulated crossover in genetic algorithms. In L. D. Whitley, editor,Fondation of Genetic Algorithms 2, pages 38-45, San Mateo, CA, Morgan Kaufmann (1993)
17. Baluja, S.: Population-based incremental learning: A method for integrating genetic search based function optimization and competitive learning. Technical Report CMU-CS-94-163, Carnegie Mellon University, Pittsburg, Pennsylvania (1994)
18. Thierens, D., Goldberg, D.: Mixing in genetic algorithms. In: Proceedings of the Fifth International Conference on Genetic Algorithms. (1993) 38–45
19. Banzhaf, W., Nordin, P., Keller, R.E., Francone, F.D.: Genetic Programming – An Introduction; On the Automatic Evolution of Computer Programs and its Applications. Morgan Kaufmann, dpunkt.verlag (1998)
20. Laszewski, G., Muhlenbein, H.: A parallel genetic algorithm for the k-way graph partitioning problem. In: 1st inter. Workshop on Parallel Problem Solving from Nature. (1990)
21. C.J. Alpert, L.W.Hagen, A.K.: A hybrid multilevel/genetic approach for circuit partitioning. Technical report, UCLA Computer Science Department (1994)
22. Hulin, M.: Circuit partitioning with genetic algorithms using a coding scheme to preserve the structure of a circuit. In: Lecture Notes in Computer Science, 496, Springer-Verlag (1989) 75–79
23. Alpert, C., Kahng, A.: Recent directions in netlist partitioning: A survey. Technical Report CA 90024-1596, UCLA Computer Science Department (1997)

24. http://vlsicad.cs.ud.edu/: (Cad benchmarking laboratory)
25. Inc, X.: Xnf: Xilinx netlist format. (www.xilinx.xom)
26. Harary, F.: Graph Theory. Addison-Wesley (1968)
27. J.I. Hidalgo, J. Lanchares, R.H.: Graph partitioning methods for multi-fpga systems and reconfigurable hardware based on genetic algorithms. In: Proceedings of the 1999 Genetic and Evolutionary Computation Conference Workshop Program. (1999) 357–358
28. Parmee, I.C., Watson, A.H.: Preliminary airframe design using co-evolutionary multi-objective genetic algorithms. In: Proceedings of the 1999 Genetic And Evolutionary Computation Conference, Morgan Kaufmann (1999) 1657–1665
29. C.A.Coello: A comprehensive survey of evolutionary-based multiobjective optimization techniques. Knowledge and Information Systems **1** (1999) 269–308
30. J.I. Hidalgo, R. Baraglia, R.P.J.L.F.T.: A parallel compact genetic algorithm for multi-fpga partitioning. In: PDP20001, 9th Euromicro Workshop on Parallel and Distributed Processing. (2001)
31. R. Baraglia, J.I. Hidalgo, R.P.: A hybrid heuristic for the traveling salesman problem. IEEE Transactions on Evolutionary Computation **5** (2001) 613–622
32. Hidalgo, J.: Multi-FPGA systems partitioning and placement techniques based on Genetic Algorithms. PhD thesis, Universidad Complutense de Madrid (2001)
33. Lohn, J.D., Colombano, S.P.: Automated analog circuit synthesis using a linear representation. Lecture Notes in Computer Science **1478** (1998) 125+
34. Miller, J.F., Job, D., Vassilev, V.K.: Principles in the evolutionary design of digital circuits-part I. Genetic Programming and Evolvable Machines **1** (2000) 7–35
35. Koza, J.R., David Andre, Bennett III, F.H., Keane, M.: Genetic Programming 3: Darwinian Invention and Problem Solving. Morgan Kaufman (1999)
36. Thompson, A., Layzell, P.J.: Evolution of robustness in an electronics design. In: ICES. (2000) 218–228
37. Seok, H.S., Lee, K.J., Zhang, B.T., Lee, D.W., Sim, K.B.: Genetic programming of process decomposition strategies for evolvable hardware. In: Proceedings of the Second NASA / DoD Workshop on Evolvable Hardware, Palo Alto, California, Jet Propulsion Laboratory, California Institute of Technology, IEEE Computer Society (2000) 25–34
38. Xilinx Corporation (www.xilinx.com)
39. Fernandez de Vega, F.: Distributed Genetic Programming Models with Application to Logic Synthesis on FPGAs. PhD thesis, University of Extremadura (2001)
40. Fernandez, F., Sanchez, J.M., Tomassini, M.: Placing and routing circuits on FPGAs by means of parallel and distributed genetic programming. In Liu, Y., Tanaka, K., Iwata, M., Higuchi, T., Yasunaga, M., eds.: Evolvable Systems: From Biology to Hardware, Proceedings of the 4th International Conference, ICES 2001. Volume 2210 of Lecture Notes in Computer Science., Tokyo, Japan, Springer-Verlag (2001) 204–214

Part III

Evolvable Designs

8

Evolutionary Computation and Parallel Processing Applied to the Design of Multilayer Perceptrons

Ana Claudia M. L. Albuquerque, Jorge D. Melo, and Adrião D. Dória Neto

Departamento de Engenharia de Computação e Automação,
Universidade Federal do Rio Grande do Norte,
Campus Universitário s/n - 59072-970 - Natal - RN, Brazil,
`(aclaudia | jdmelo | adriao)@dca.ufrn.br`

Neural networks consist of very powerful tools and had their use extended vastly due to their ability of providing great results to a broad range of applications. The combination of evolutionary computation, such as genetic algorithms and parallel processing can be very powerful when applied to the learning process of the neural network, as well as to the definition of its architecture. A lot of research has been developed combining and applying evolutionary computation into the design of neural networks. It is very important to emphasize that most of the learning algorithms developed to train neural networks only refine their synaptic weights, not considering the design of the networks architecture. However, it is a very hard task to define the neural networks architecture for specific applications under given sets of constraints. To a large extent, that could be a process of trial and error, relying mostly on past experience with similar applications. Evolutionary algorithms, on the other hand, offer attractive ways to search for optimal solutions in a variety of problem domains. Due to this characteristic, the definition of architectures for neural networks becomes a natural candidate for the application of evolutionary algorithms, such as genetic. Also, the learning process of neural networks can be very slow which can put in danger the performance of countless applications. Therefore, the use of parallel processing is essential in minimizing the time required on the training process, improving the applications performance. Furthermore, the use of cooperation in the genetic algorithm allows the interaction of different populations, avoiding local minima and helping in the search of the ideal solution, accelerating the evolutionary process. Finally, individuals and evolution behavior can be exclusive on each copy of the genetic algorithm running in each task enhancing the diversity of populations.

8.1 Introduction

The use of neural networks on the solution of problems has become even more usual due to the great results provided by these powerful tools. More specifically, the Multilayer Perceptron neural networks have received a lot of attention due to their desirable characteristics such as versatility, simplicity, computational efficiency, accuracy and high degree of applicability, which have motivated the use of these tools as a global interpolator and as a pattern classifier [1].

However, it is well known that, in general, the design of artificial neural networks is a very hard task. Defining architectures of neural networks for specific applications is, therefore, basically a process of trial and error, relying mostly on past experience with similar applications [2].

The great majority of learning algorithms designed to train neural networks, such as the error back-propagation algorithm, are not able to determine an ideal architecture for a certain application. Instead, they only refine the network's synaptic weights.

Thus, techniques for automating the design of neural networks are clearly of interest and a natural candidate for the application of evolutionary algorithms.

A lot of research has been developed combining and applying evolutionary computation to the design of neural networks [3]. Therefore, instead of using the classical error back-propagation algorithm in the learning process of the neural network, a different approach using genetic algorithms to train the neural network and to define its architecture is introduced in this chapter.

Neural networks and genetic algorithms can be combined in a way that a population of neural networks competes against each other in a Darwinian setting. Each individual of the population represents a certain neural network with differing architecture and synaptic weights. The individuals codifying neural networks that produce good results are combined and passed onto the next generation. After a number of iterations, an optimized neural network can be obtained.

The learning process of the Multilayer Perceptron neural network can be very slow, varying according to the size of the network, which can put in danger the performance of countless applications. Aiming to minimize the time required on the training process and to improve the applications performance, the use of parallel processing and techniques were incorporated into the training algorithm developed.

The main concept of parallelism consists of dividing and executing a large number of tasks simultaneously. Therefore, knowing that the essence of genetic algorithms consists of having populations fighting against each other in search of the fit solution, it is very easy to see that there are many ways to explore parallelism in genetic algorithms, as it will be introduced later.

Several parallel processes were created, each one of them corresponding to a different population. All populations created will evolve simultaneously. At the end of a predetermined number of generations, the created processes will be able to communicate in order to exchange information concerning the best individuals selected by each one of them. That procedure goes on until the error signal produced by the neural network tends to zero and, consequently, the network can be considered trained.

The exchange of information between the populations is very important, allowing them to cooperate and exploit promising areas of the search space found by other populations and, also, reintroduce in the population previously lost genetic material [4].

Furthermore, different reproduction and evolution behaviors were introduced in each one of the coexisting populations. The use of distinct evolution behavior will contribute on the maintenance of the diversity of the individuals regarding each population.

This chapter is organized in the following way. In Section 8.1, an introductory view of the combination of genetic algorithms and parallel processing into the training process of Multilayer Perceptrons neural networks is given. In Section 8.2 and Section 8.3, it will be presented an overview of neural networks and Multilayer Perceptrons, respectively. In Section 8.4, it will be explained, with further details, the use of genetic algorithms and parallel processing in Multilayer Perceptrons. In Section 8.5, it will be presented the use of genetic algorithms in defining the neural network's architecture and in refining its synaptic weights. In Section 8.6, a different approach of a cooperative parallel genetic algorithm with different evolution behaviors is given. In Section 8.7, applications on approximation of functions will be illustrated. Finally, in Section 8.8, it will be presented the conclusions obtained by the combination of genetic algorithms and parallel processing into the design of Multilayer Perceptrons neural networks.

8.2 Artificial Neural Networks

Artificial neural networks consist of an architecture projected to simulate the way as the brain accomplishes a certain task. It is composed by processing units, denominated artificial neurons, which introduce the capacity of storing experimental knowledge in order to be available for the practical use. Basically, an artificial neural network resembles the human brain in the following aspects:

- The network, starting from its environment, acquires knowledge through a learning process;
- The acquired knowledge is stored in the connections within the neurons, known as synaptic weights.

The artificial neurons are connected by communication channels that are associated to a certain synaptic weight and only operate their local data, which are inputs received by their connections. The intelligent behavior of the neural network comes from the interactions between these several processing units of the network.

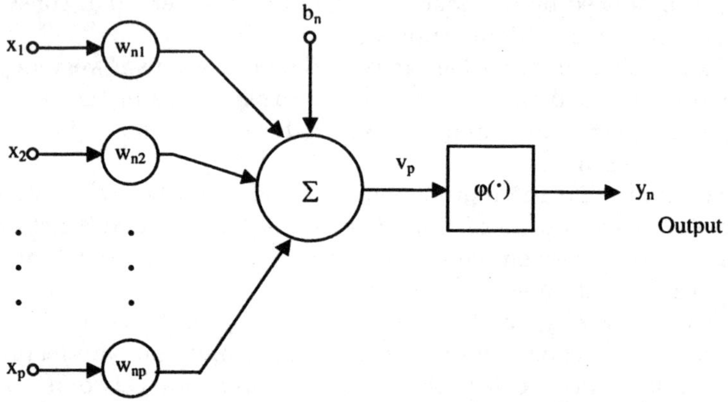

Fig. 8.1. Artificial neuron

As it can be observed in Fig. 8.1, the artificial neuron presents a set of synapses, each one characterized by a self-weight (w_{ij}). An input signal x_p in the entrance of the *p-th* synapse connected to the neuron n is multiplied by the synaptic weight w_{np}. The summation symbol, represented by the Greek letter Σ in Fig. 8.1, is used to add the input signals, weighted by the respective synapses of the neuron. The activation function, represented in Fig. 8.1 by $\varphi(.)$, is used to restrict the output amplitude of the neuron. This function is going to restrict the allowed range of the output amplitude signal to a finite value. Typically, the normalized range of the output amplitude of a neuron is written as the closed unitary interval [0,1], or as the closed interval [-1,1]. The bias, represented by b_n, has the effect either to increasing or to decrease the network input of the activation function, depending whether it is positive or negative, respectively.

The learning algorithms, used in the training process of artificial neural networks, basically modify the synaptic weights of the network in an ordered way until there is the production of a wished output. Thus, it is said that an artificial neural network is trained for a given problem when it produces an equal or nearly equal response to the desired one. As previously mentioned, this training procedure can be very slow, depending, mostly, on the complexity of the neural network's architecture (a large number of layers as well as of neurons per layers). Therefore, a good alternative to improve this performance and, consequently, to decrease the time of training, would be the application of parallelism techniques in the training algorithm of the neural network.

8.3 The Multilayer Perceptron Neural Network

The Multilayer Perceptron neural networks belong to an important class of neural networks. They were developed for the resolution of more complex problems, which could not be solved by using the model of basic neuron proposed by Rosemblatt in [5], since this model works properly only regarding to problems that are linearly separable. For example, a sole perceptron or a combination of the outputs of some perceptrons, would not be able to learn a logic or exclusive operation (XOR), once it defines a non-linear problem. To do that it will be necessary to introduce more connections, which exist only in a perceptron network disposed in layers. It is worth to point out the importance of these internal neurons in the neural network, once it was proved that without the presence of such units the resolution of linearly not separable problems would be impossible. Thus, the Multilayer Perceptron neural networks are constituted by a set of sensor units forming the input layer, one or more hidden layers, and an output layer of computational nodes. Fig. 8.2 illustrates the architecture of the Multilayer Perceptron.

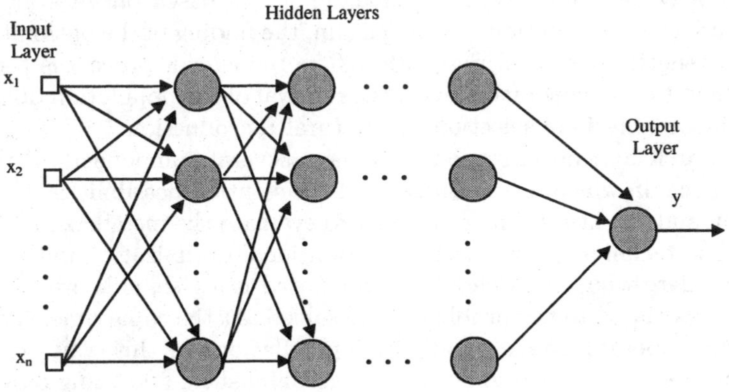

Fig. 8.2. Organization in layers of the Multilayer Perceptron

A Multilayer Perceptron neural network has three distinctive characteristics [1]:

1. The model of each neuron of the network includes a smooth non-linear activation function. In other words, it is differentiable at any point, since it does not present the abrupt limitation used in Rosenblatt's perceptron.
2. The network contains one or more layers of hidden neurons, which are part of the input or output of the net. These hidden neurons enable the network to learn complex tasks extracting progressively the most significant characteristics in the standard (vector) inputs.

3. The network exhibits a high degree of connectivity, determined by its synapses. A modification in the connectivity of the network requires a change in the population of the synaptic connections or of their weights.

It is through the combination of these characteristics, together with the ability to learn the experience through the training, that the Multilayer Perceptron neural network derives its computational power.

8.4 Genetic Algorithms and Parallelism in Multilayer Perceptron Learning

The genetic algorithms belong to one of the five areas of the evolutionary computation that is based on the selection theory and the natural evolution. Thus, it is proposed a model of computational structures that evolve with the goal of improving the general performance of the population regarding a set of individual characteristics. Such characteristic, then, translate the adaptation or individual's adequacy to the environment. Therefore, the genetic algorithms consist in dynamic methods of search based on selection mechanisms and natural evolution having, as aim, the finding of the optimal individual of a genetically refined population. The refinement process is then given from generation to generation, with the renewal of the population obeying the probabilistic criteria of selection and natural reproduction.

All search and optimization task have several components such as the search space, in which are considered all the solution possibilities of a certain problem, and the fitness function, used to evaluate the members of the search space. The techniques for search and traditional optimization initiate with a sole candidate who, iteratively, is manipulated using some heuristics (static) directly associated to the problem to be solved. At the same time, the genetic algorithms operate over a candidates' population. In this way, searches in different areas of the solution space are accomplished, allocating a number of members appropriated for the search in several regions.

Much of conventional methods of maximization or minimization from certain characteristics of a given individual, use deterministic criteria to move from a point to the other in the search hyperspace. However, being a multimodal function, i.e., a function with several peaks within a same interval, these methods can result in a premature stop or in the paralysis of the search process in a maximum (or minimum) local point, instead of a maximum (or minimum) global point. It is possible, however, to overcome this problem producing a new population to each iteration or generation allowing, thus, the simultaneous exploration of several points of the hyperspace. In this way, many maximum (or minimum) can be explored efficiently reducing, consequently, eventual stops in undesirable maximum (or minimum) locals.

To code the characteristics involved in the optimization process, chains of the binary alphabet are frequently used. During the execution phase, then, the length of the chain usually remains fixed, depending on the degree of precision, required for the solution of the problem, or of the amount of characteristics in observation.

As stated previously, parallel processing techniques have been used in the learning process of the Multilayer Perceptron networks in order to minimize the amount of time consumed on the training process of more complex networks architectures and enlarge their range of applications. Several possibilities have been exploited in literature [6], [7]. In [8] a new parallel algorithm was presented, based on the cooperation concept. Multiple copies of the neural network in each task allow new parallel strategies. Information is periodically exchanged among the tasks to efficiently guide the search procedure and the back-propagation algorithm was used to refine the connection weights and minimize the error signal.

The use of parallel genetic algorithms is a very good alternative since they are able to improve the performance of the genetic algorithms both in terms of velocity as in terms of enhancing the search quality, allowing, thus, simultaneous exploration of several points of the search space. Consequently, the maintenance of more diverse subpopulations helps to avoid premature convergence (local minima).

In genetic algorithms, parallelism was exploited in different levels producing both coarse and fine grain solutions. The fine-grained model assumes the placement of only one member of the population on each processing node. Therefore, the individuals can only reproduce and exchange genetic material with other individuals in a bounded region, as opposed to global ones. The coarse-grain model, on the other hand, is the most popular model used and assumes the division of a large population into several subpopulations with or without communication among the tasks [9]. Therefore, multiple processors run a sequential genetic algorithm on their population. Cooperation can be used in coarse-grain parallel genetic algorithms, where processors exchange individuals from their subpopulation with other processors. In the island model, individuals can randomly migrate from one population to another. In the stepping stone model, however, the individuals can migrate only to geographically nearby subpopulations. The existence of isolated subpopulations will help in the maintenance of genetic diversity.

Also, the concept of cooperative parallel algorithms differs from the traditional approach of parallel algorithm development [10]. Instead of dividing the computational load among the tasks and having them only to compete against each other, complete problems are solved in each task (competitive approach) and their solutions are speeded-up with the information provided by the other ones (cooperative approach).

Individuals and evolution behavior can be exclusive on each copy of the genetic algorithm running in each task and communications can be used to accelerate the evolutionary process. In the following sections, this approach will be exploited.

8.5 Training Multilayer Perceptron with Genetic Algorithms

A genetic algorithm consists of a dynamic search method based on the theory of natural selection and can be successfully used in the learning process of a Multilayer Perceptron neural network since each individual of the population is represented by a finite string of binary symbols, known as chromosome [11]. The chromosome of an individual will encode one specific neural network. Therefore, in each population there are a finite number of individuals representing one neural network, with differing architecture, as well as synaptic weights. At the end of the training process, the selected individual will hold the final configuration of the neural network.

The genetic algorithm will proceed in the following way: an initial population of individuals is generated randomly. Each individual's genetic material, which consists of a vector of 0s and 1s, will contain a description of a specific neural network. Later, the individuals are decoded and evaluated according to a predefined fitness function. The best individuals are the ones capable to approximate the fitness function to zero.

The fitness function, however, is calculated for each set of the training samples by adding all the mean square error signals obtained for each element located in the output of the neural network and dividing the result by the number of samples applied to the network. The error signal, meanwhile, is the result of the subtraction between a desired response previously defined and the actual response obtained by the network.

If the stop condition is not yet reached, the learning process continues and a certain number of individuals are chosen, according to the selection criteria, to be the parents of the next generation. In order to maintain a high degree of diversity among the individuals, different reproduction criteria were incorporated into the genetic algorithm. In general, it can be said that to form a new population, individuals are selected according to their capability of minimizing the fitness function. Thus, the individuals that obtained the smaller errors signals have a better chance of reproducing, while the others are more likely to disappear.

8.6 Cooperative Parallel Genetic Algorithm with Different Evolution Behaviors

The cooperative parallel genetic algorithm with different evolution behaviors developed is used to determine the Multilayer Perceptron neural network synaptic weights and architecture. In each population, there are a finite number of individuals codifying one specific Multilayer Perceptron neural network with differing architectures and synaptic weights. As the populations evolve, the network's synaptic weights are being refined along with its architecture.

In the end, from the genetic material of the individual that best fit the solution, both architecture and set of synaptic weights for the neural network are extracted.

It is worth to stress that the definition of neural networks architectures can become a process of trial and error, relying mostly on past experience with similar applications. Also, the performance of neural networks on a large number of applications is critically dependent on the choice of an ideal architecture. As a result, it is very hard to pre-define architectures for Multilayer Perceptron neural networks for a certain problem without previous knowledge or experience with similar applications. The use of genetic algorithms is, therefore, a natural and intuitive way to accomplish such task.

Besides defining the network's architecture, the genetic algorithm is used, simultaneously, to refine the network's synaptic weights. In order to enhance the search for the fit individual, different reproduction criteria were incorporated into the genetic algorithm. The use of different reproduction criteria will contribute on the maintenance of the diversity of the individuals regarding each population, speeding up the search for the ideal solution. Therefore, existing populations will evolve differently from one another following its own criterion of reproduction.

The reproduction step of a genetic algorithm can be performed using many different kinds of heuristics. In the algorithm developed, the individuals are ordered according to the value of the fitness function produced by them. The quantity of individuals that will become parents of the next generation is, then, chosen randomly. Note that the parents of the next generation are the individuals that produced the lowest error signals. From the set of individuals chosen to be parents of the next generation, two are randomly chosen to have its genetic material combined, producing an offspring. The genetic material can be combined in one or two different points. A small number of mutations, also obtained in a random manner, were introduced to the new population. Fig. 8.3 illustrates two examples of reproduction of the genetic algorithm.

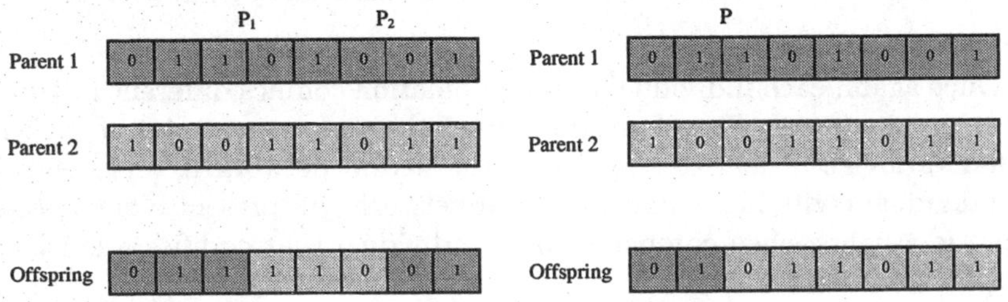

Fig. 8.3. Illustration of two examples of reproduction of the genetic algorithm

Besides presenting different evolution behaviors, the existing populations are able to communicate with each other and exchange valuable information on the best individual selected by each one of them so far. The presence of cooperation in the genetic algorithm is, therefore, very important since the exchange of information between the populations helps to avoid local minima. Also, it allows the exploitation of a larger range of the search space and reintroduces previously lost genetic material.

The genetic material of each individual contains two main fields. The first field codifies an index to a table of architectures. The second field codifies the synaptic weights for the architecture defined in the first field. Each synaptic weight is represented by a 32-bits binary string. The index to a table of architectures is also represented by a 32-bits binary string. During the initialization, the binary strings representing the networks architecture and synaptic weights are generated randomly. As the populations evolve, the best architectures and set of synaptic weights are maintained. Finally, the best individual will hold the final configuration of the neural network. Fig. 8.4 illustrates the representation of the genetic material of each individual. Note that the table of architectures will contain different kinds of networks with different numbers of layers, as well as neuron per layer. However, the input and output dimensions of the neural network are pre-determined since they depend and vary according to the application.

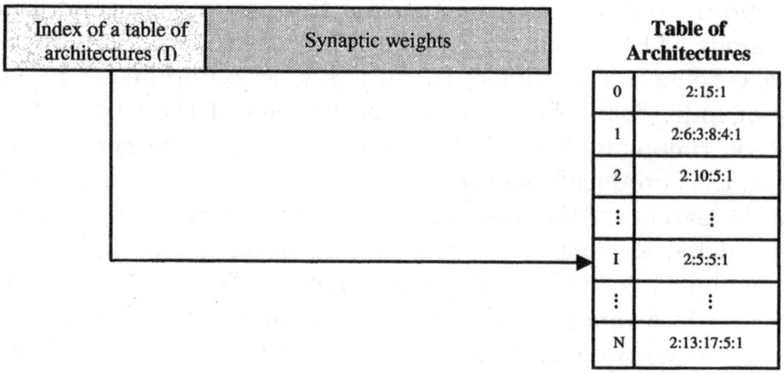

Fig. 8.4. Representation of the genetic material of an individual

Once again, each individual of the population codifies different neural networks architectures. Therefore, the size of the genetic material of each individual varies according to the size of the neural network it represents, i.e., an individual codifying a simpler neural network will present a smaller set of synaptic weights when compared to an individual that codifies a much more complex architecture.

However, it is necessary to standardize the representation of the genetic materials since, performing operations on differently sized chromosomes, memory that is not part of the smaller chromosome will be used in the recombination process.

The simpler solution to this problem consists on taking as standard the size of the genetic material of the individual that codifies the largest neural network. As a result, all individuals would present chromosomes with same sizes and the layers, neurons and connections that do not exist would be represented by zero. However, by doing so, the size of the neural networks and the genetic materials could increase inappropriately. Besides demanding a lot of memory and making the recombination process very slow, this solution increases the amount of time required on the exchange of messages performed by the parallel tasks, which can badly affect the final performance of the algorithm.

The most viable solution found was based on the specification of a reasonable interval of synaptic weights for the networks. Therefore, all possible neural networks architectures that obeyed the specified interval were considered. By doing so, the uncontrolled growth of the genetic materials can be avoided. Furthermore, the massive use of memory space is also avoided and, more importantly, the message exchange among parallel tasks is accelerated. Therefore, during the initialization of the populations, the amount of individuals at each population and the interval of synaptic weights are previously specified. Later, the genetic material of each individual is generated randomly.

In Fig. 8.5, the parallel structure adopted is illustrated.

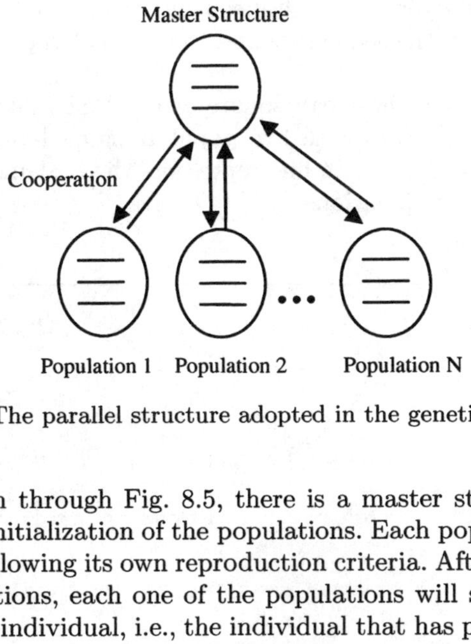

Fig. 8.5. The parallel structure adopted in the genetic algorithm

As can be seen through Fig. 8.5, there is a master structure that is responsible for the initialization of the populations. Each population will evolve simultaneously, following its own reproduction criteria. After a predetermined number of generations, each one of the populations will send to the master structure the best individual, i.e., the individual that has produced the lowest error signal. From this set of best individuals received from all the coexisting populations, the master structure will select the one with the lowest error signal. Finally the master structure will send a copy of the individual selected to each one of the coexisting populations and so on, until the error signal tends to zero.

The environment used for the implementation of the parallel algorithm was the PVM (Parallel Virtual Machine). The software PVM is an environment for parallel and distributed computation. It allows the user to create and to access a parallel computation system made up from a collection of distributed processes, as well as to treat the resultant system as a unique virtual machine, hence the name parallel virtual machine. The software PVM is based on the message-exchange parallel programming model. In this way, messages are exchanged among the tasks through a connection chain.

8.7 Application on Approximation of Functions

The genetic algorithm developed in this chapter was used in the learning process of the Multilayer Perceptron network in order to approximate functions. The first two results illustrate the approximation of the functions $f(x) = 1/x$ and $f(x) = sin(2\pi x)$. Later, two more complex functions, $z = sin(r)/r$ and $f(x1, x2) = cos(2\pi x1).cos(2\pi x2)$, were used.

As was said before, the great advantage of the cooperative parallel genetic algorithm is the concept of having different populations with different evolutional behaviors evolving simultaneously. To illustrate that, it is drawn a comparison between the sequential and the parallel approach of genetic algorithms.

Fig. 8.6 represents the mean square error (MSE) signal of the usual sequential genetic algorithm with five populations evolving independently for the function $f(x) = 1/x$. Each line represents the best individual selected by each one of the five populations.

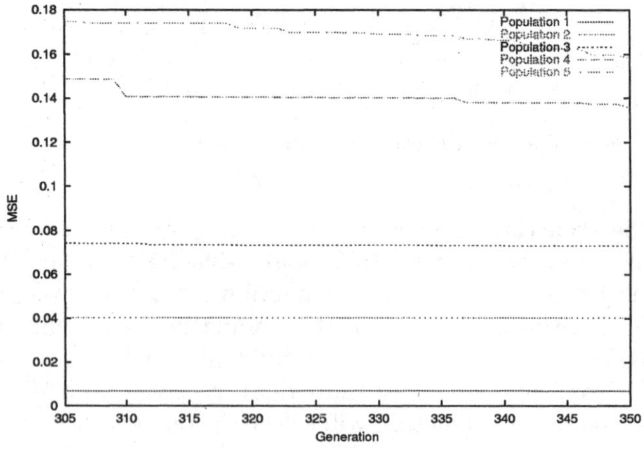

Fig. 8.6. MSE signal of the sequential genetic algorithm for the function $f(x) = 1/x$

Therefore, as the populations evolve independently, this is equivalent to an isolated genetic algorithm being executed, i.e., there is one population and new generations are born from it, always aiming to find the individual that best fits the solution. No exchange of information or cooperation exists among the populations.

According to the initialization, is very easy to find populations stuck in local minima. For instance, the members of Population 1, 2 and 3 seem to have its MSE signal stabilized. On the other hand, the members of Population 4 and 5 are slowly reducing the MSE signal but were unfortunate during the initialization, presenting the two highest error signals produced.

On the contrary, through Fig. 8.7, where there are five populations evolving simultaneously, it is possible to realize the improvement that the cooperative parallel genetic algorithm with different evolution behavior represents.

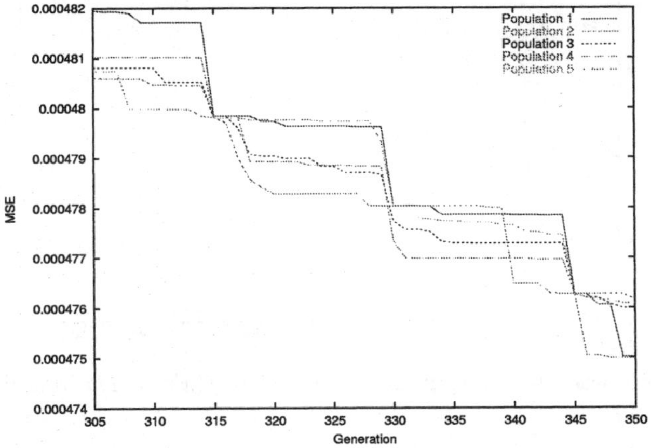

Fig. 8.7. MSE signal of the cooperative parallel genetic algorithm for the function $f(x) = 1/x$

As said before, the five populations appear evolving in independent ways, exchanging genetic material once in a while. On the 315th generation, for instance, the members of Population 2 contain the best of all the individuals, i.e., the one with the lowest MSE signal. On the 330th generation, all tasks communicate, meaning that all populations will receive a copy of the best individual produced so far. By doing so, we allow that all the populations that initially evolved independently have a common aspect: the fit individual. From then on, the populations continue evolving independently again. Note that due to the communication during the 330th generation, the MSE signal produced by all populations dropped, especially the one produced by Population 4 that accomplished the lowest MSE signal right after this episode.

The parallel genetic algorithm developed was also used to define the Multilayer Perceptron neural network architecture. Several configurations of neural networks were used. The number of hidden layers varied from one to six and the number of neurons per hidden layer varied from three to fifteen. There was one neuron in the input layer and one neuron in the output layer. The number of neurons per input layer and the number of neurons per output layer were pre-determined since they depend on the application.

The final configuration of the Multilayer Perceptron neural network presented one neuron in the input layer, one neuron in the output layer and one hidden layer with ten neurons.

In Fig. 8.8 is illustrated the reconstructed output of the Multilayer Perceptron along with the original function.

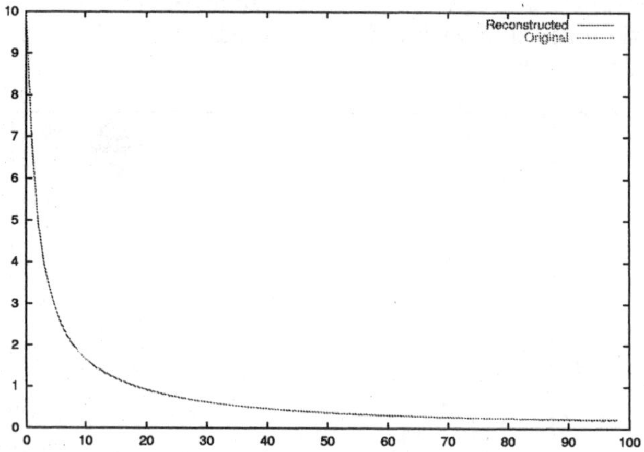

Fig. 8.8. Reconstructed output of the function $f(x) = 1/x$ obtained from the Multilayer Perceptron

The results obtained by the approximation of the function $f(x) = sin(2\pi x)$ are now presented.

First, a sequential genetic algorithm is used and the MSE signal of the five populations evolving independently is illustrated on Fig. 8.9.

As can be seen through Fig. 8.9, there are five populations evolving, without any cooperation or communication. The new generations are born from its population only.

Once again, it can be seen that without the cooperation, the populations are more likely to get stuck in local minima. On this example, all the populations are reducing the MSE signals very slowly, which can badly affect the performance of the neural network.

In Fig. 8.10, on the other hand, there are five populations evolving simultaneously by the use of the cooperative parallel algorithm with different evolution behaviors. Once again, the use of different behaviors during the reproduction phase of the genetic algorithm was crucial in helping maintaining the diversity of the populations, fastening the obtainment of convergence.

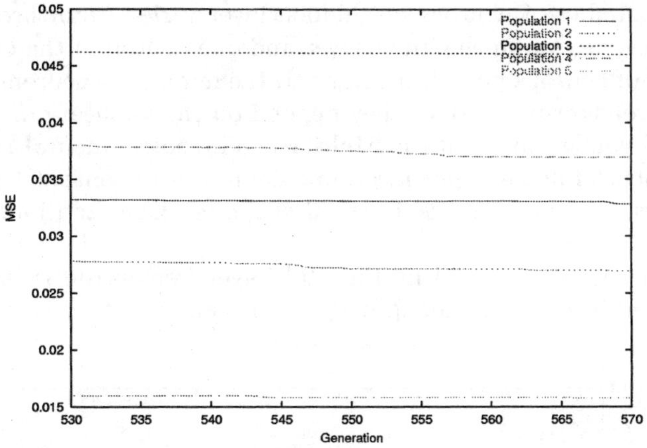

Fig. 8.9. MSE signal of the sequential genetic algorithm for the function $f(x) = sin(2\pi x)$

For instance, from the 540th generation up to the 555th generation, the individuals from Populations 1, 3, 4 and 5 presented a higher MSE signal than the ones from Population 2. During the 555th generation, by means of communication, the genetic material of all populations was exchanged and all the MSE signals were brought down to the same level.

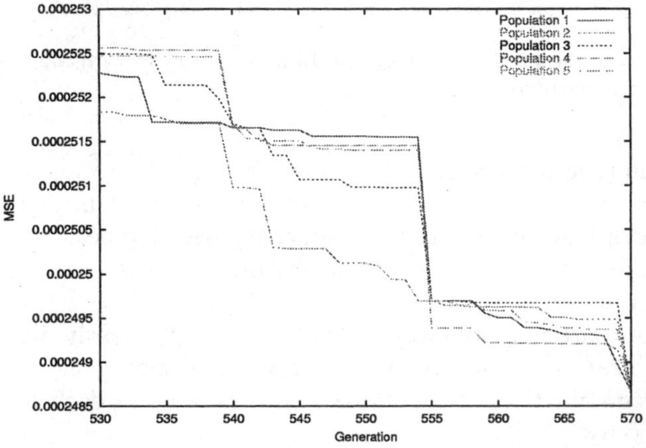

Fig. 8.10. MSE signal of the cooperative parallel genetic algorithm for the function $f(x) = sin(2\pi x)$

The Multilayer Perceptron neural network architecture was defined by the use of the parallel genetic algorithm developed. Several configurations of neural networks were used. The number of hidden layers varied from one to six and the number of neurons per hidden layer varied from three to fifteen. There was one neuron in the input layer and one neuron in the output layer. The number of neurons per input layer and the number of neurons per output layer were pre-determined since they depend on the application.

The final configuration of the Multilayer Perceptron neural network presented one neuron in the input layer, one neuron in the output layer and two hidden layers with five neurons in the first hidden layer and four neurons in the second.

The final output obtained by the Multilayer Perceptron is reconstructed and illustrated in Fig. 8.11, along with the original function.

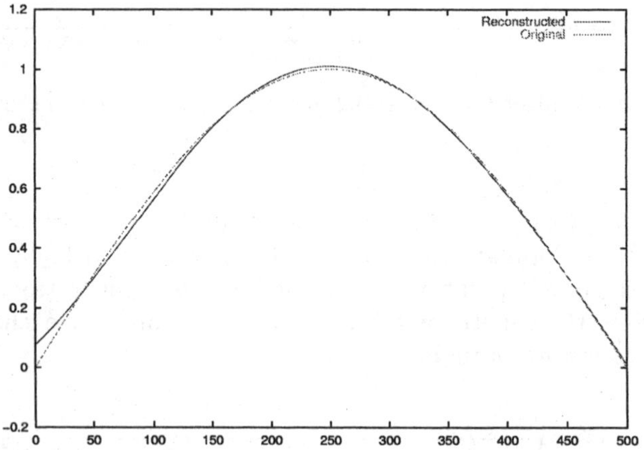

Fig. 8.11. Reconstructed output of the function $f(x) = sin(2\pi x)$ obtained from the Multilayer Perceptron

Later, the function $z = sin(r)/r$, where $r = \sqrt{x^2 + y^2}$ and -7.5 $<= x <=$ 7.5 and -7.5 $<= y <=$ 7.5 was approximated by the Multilayer Perceptron neural network trained with the cooperative parallel genetic algorithm.

Once again, a sequential genetic algorithm was first used on the approximation of the function.

Thus, there are five populations evolving independently, without any degree of cooperation neither communication. The new generations are born from the recombination of individuals that belong to each one of the five populations only.

The MSE signal of the five populations is illustrated on Fig. 8.12. However, it can also be noticed the existence of populations, such as 1 and 3, presenting higher error signals that, due to the lack of cooperation, are not helped by the others. The performance of the neural network can, then, be affected in a bad way.

Later, the cooperative parallel genetic algorithm with different evolution behaviors was used on the approximation of the function $z = sin(r)/r$. The

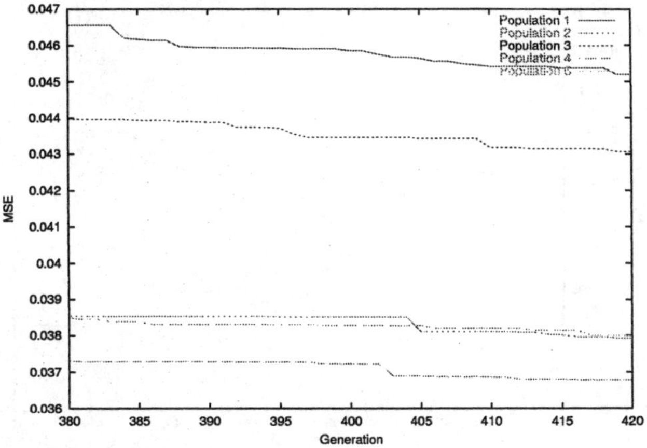

Fig. 8.12. MSE signal of the sequential genetic algorithm for the function $z = sin(r)/r$

use of parallel processing, along with cooperation and different evolution behaviors, once again helped to accelerate the convergence of the neural network.

In Fig. 8.13, there are five populations evolving simultaneously, where each line represents the best individual produced so far by each one of them.

Right from the start, it can be noticed, through Fig. 8.13, that the MSE signals of populations 1, 2, 3 and 5 were brought down to the same level of the Population 4 at the 360th generation. Then, all the existing populations went on evolving simultaneously, competing and cooperating with one another every once in a while. Finally, from the help of its neighbors, the Population 2, which started out with the second highest MSE signal, accomplished the lowest one in the end, at generation 420.

Once again, the parallel genetic algorithm developed was used to define the Multilayer Perceptron neural network architecture. Several configurations of neural networks were used. The number of hidden layers varied from one to six and the number of neurons per hidden layer varied from three to fifteen. There were two neurons in the input layer and one neuron in the output layer. The number of neurons per input layer and the number of neurons per output layer were pre-determined since they depend on the application.

The final configuration of the Multilayer Perceptron neural network presented two neurons in the input layer, one neuron in the output layer and two hidden layers with ten neurons in the first hidden layer and six neurons in the second.

The reconstructed output of the function $z = sin(r)/r$ is illustrated in Fig. 8.14. In Fig. 8.15, it is presented the original output of the function.

Finally, the Multilayer Perceptron neural network was used to approximate the function $f(x1, x2) = cos(2\pi x1).cos(2\pi x2)$.

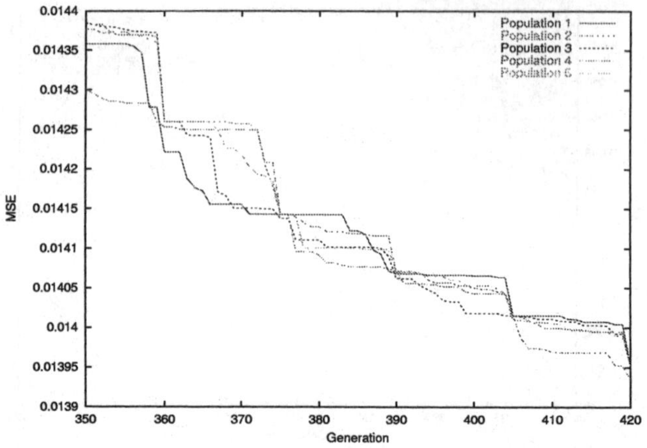

Fig. 8.13. MSE signal of the cooperative parallel genetic algorithm for the function $z = sin(r)/r$

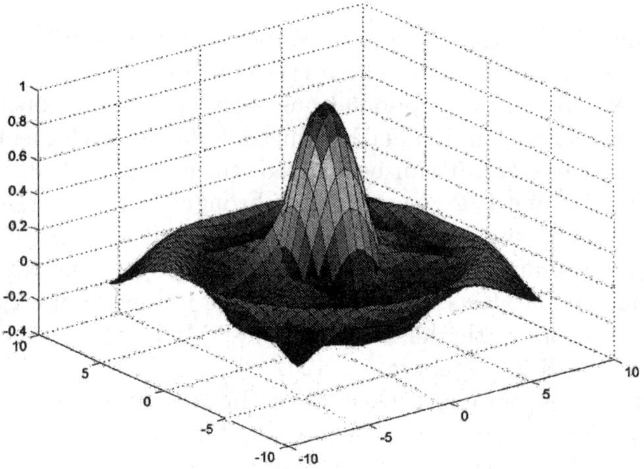

Fig. 8.14. Reconstructed output of the function $z = sin(r)/r$ obtained from the Multilayer Perceptron

First, a sequential genetic algorithm is used and the MSE signal of the five populations evolving independently is illustrated on Fig. 8.16.

All five populations evolve independently, without exchange of information or cooperation among them. Once again, with this approach, is very easy to find populations stuck in local minima, such as Populations 3, 4 and 5.

On the contrary, through Fig. 8.17, where there are five populations evolving simultaneously, it is possible to realize the improvement that the cooperative parallel genetic algorithm with different evolution behavior represents.

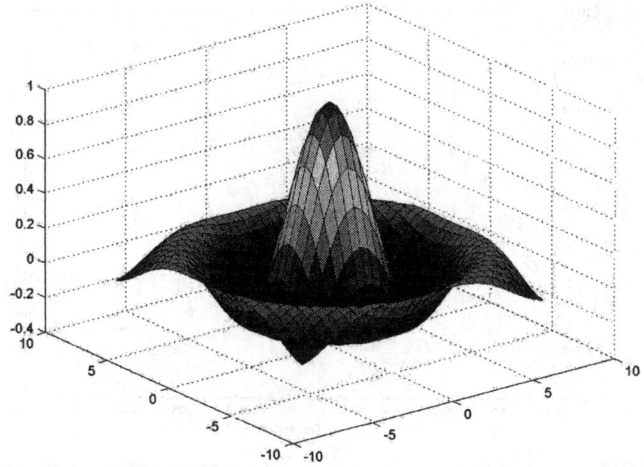

Fig. 8.15. Original output of the function $z = sin(r)/r$

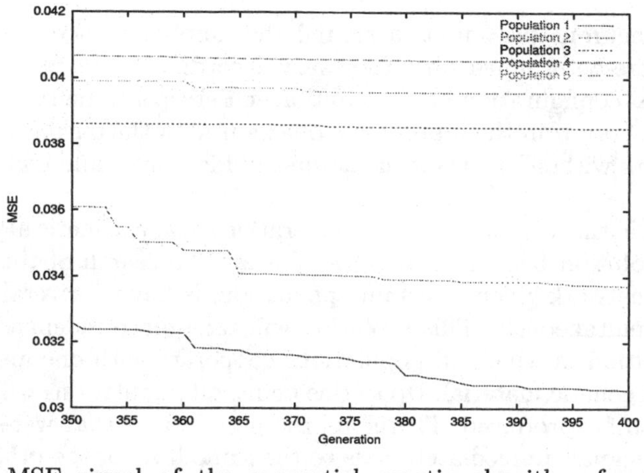

Fig. 8.16. MSE signal of the sequential genetic algorithm for the function $f(x1, x2) = cos(2\pi x1).cos(2\pi x2)$

As can be seen through Fig. 8.17, from the 360th generation up to the 375th, the members of Population 4 were the ones with the lowest MSE signal. However, at the 375th generation, due to the exchange of genetic material among the populations, the MSE signal produced by Populations 1, 2, 3 and 5 were brought down to the same level of Population 4.

The parallel genetic algorithm developed was also used to define the Multi-layer Perceptron neural network architecture. Several configurations of neural networks were used. The number of hidden layers varied from one to six and the number of neurons per hidden layer varied from three to fifteen. There were two neurons in the input layer and one neuron in the output layer. The

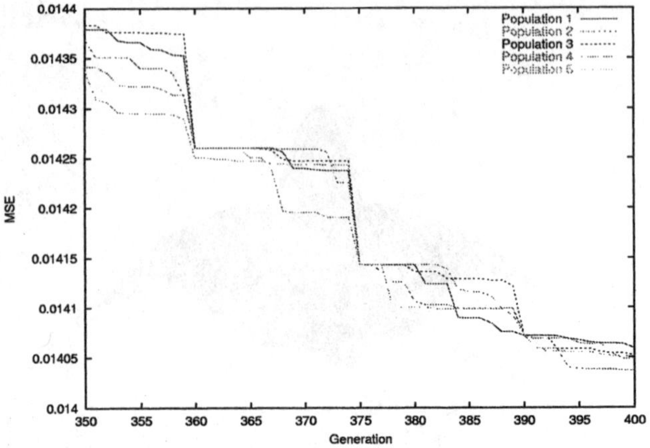

Fig. 8.17. MSE signal of the cooperative parallel genetic algorithm for the function $f(x1, x2) = cos(2\pi x1).cos(2\pi x2)$

number of neurons per input layer and the number of neurons per output layer were pre-determined since they vary according to the application.

The final configuration of the Multilayer Perceptron neural network presented two neurons in the input layer, one neuron in the output layer and two hidden layers with six neurons in the first hidden layer and eight neurons in the second.

Therefore, the advantage of the cooperative parallel genetic algorithm with different evolution behaviors, besides allowing the search of the ideal architecture for network given a certain application, is having several populations evolving simultaneously. This evolution will take place independently, up to a given position in which all populations cooperate with one another by the exchange of genetic material. Up to this point, all populations will have the fit individual so far produced. Therefore, the populations that were captured in high errors signals immediately pass to the same level of the others, avoiding local minima. In other words, while the existence of differing evolution behaviors helps maintain the diversity, the use of cooperation gives the opportunity to all the populations to evolve in such a way that they will become the best one.

The reconstructed output of the function $f(x1, x2) = cos(2\pi x1).cos(2\pi x2)$ is illustrated in Fig. 8.18. In Fig. 8.19, it is presented the original output of the function.

8.8 Summary

The cooperative parallel genetic algorithm with different evolution behaviors used in the learning process of the Multilayer Perceptron neural network was

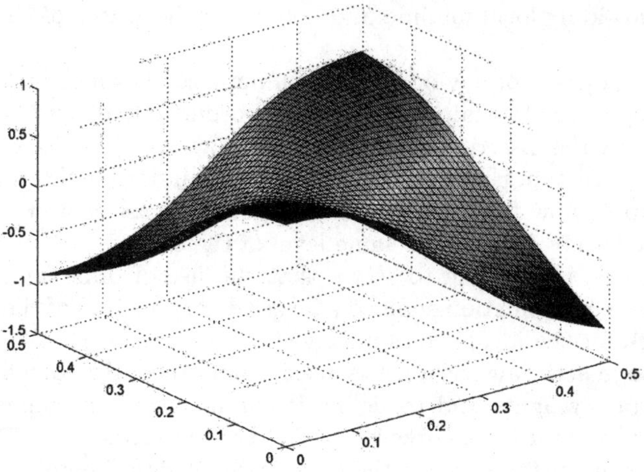

Fig. 8.18. Reconstructed output of the function $f(x1, x2) = cos(2\pi x1).cos(2\pi x2)$ obtained from the Multilayer Perceptron

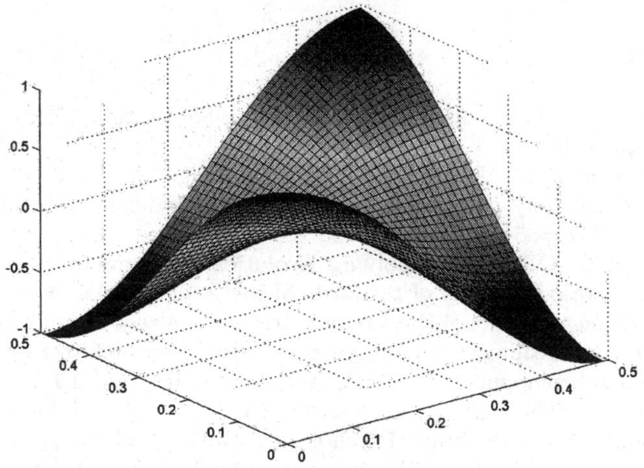

Fig. 8.19. Original output of the function $f(x1, x2) = cos(2\pi x1).cos(2\pi x2)$

very efficient when compared to the sequential form of the genetic algorithm. The cooperative parallel genetic algorithm was applied to the approximation of functions but can have its use extended to the several other kinds of applications with neural networks.

As it could be seen through the analysis of the obtained results, the simple fact of allowing the exchange of information among different populations represented a great achievement in the final performance of the algorithm. Therefore, through cooperation, different populations could interact within each other, avoiding local minima and helping in the search of the ideal solution.

Thus, the concept of having a pure competitive algorithm, where populations will only compete against each other in order to find the best individual was modified by the insertion of cooperation between populations. In this way, starting from a certain position, the different populations will contain the best individual up to now found. Therefore, populations associated to high error signals immediately drop to the same level of the others.

Furthermore, it is worth to stress that the use of different evolutionary behaviors in each population enabled a larger diversification of them, speeding up the search for the ideal solution as well.

Also, the genetic algorithm was used simultaneously to refine the network's synaptic weights and to define its architecture. In general, defining architectures for neural networks is a very hard task and most of the learning algorithms developed only refine their synaptic weights. Therefore, the combination and application of evolutionary computation into the design of neural networks are clearly of interest.

References

1. S. Haykin, Neural Networks: A Comprehensive Foundation, Maxwell Macmillan International, 1994.
2. K. Balakrishnan, V. Honavar, Evolutionary Design of Neural Network A Preliminary Taxonomy and Guide to Literature, Artificial Intelligence Research Group, CS TR #95-01, January 1995.
3. D. Curran, C. ORiordan, Applying Evolutionary Computation to Designing Neural Networks: A Study of the State of the Art, Technical Report: NUIG-IT 111002, 12 pages, National University of Ireland, Galway, Ireland, October 2002.
4. M. Potter, K. DeJong, Cooperative coevolution: An architecture for evolving coadapted subcomponents, Evolutionary Computation, 8(1):1–29, 2000.
5. F. Rosenblatt, The Perceptron: A probabilistic model for information storage and organization in the brain, Psychological Review, vol. 65. pp. 386-408, 1958.
6. J. Torresen, Parallelization of Back-propagation Training for Feed-Forward Neural Networks, PhD Thesis, The Norwegian Institute of Technology, 1996.
7. S. Foo, P. Saratchandran, N. Sundararajan, Parallel Implementation of Back-propagation Neural Networks on a Heterogeneous Array of Transputers, IEEE Trans. Systems, Man and Cybernetics - Part B, Vol. 27:1, pp. 118-126, 1997.
8. R. Alves, J. Melo, A. Dória Neto, A. Albuquerque, New Parallel Algorithms for Back-Propagation Learning, INNS-IEEE International Joint Conference on Neural Networks, Honolulu, USA, 2002.
9. P. Adamidis, S. Kazarlis, V. Petridis, Advanced Methods for Evolutionary Optimisation, LSS'98, 8th IFAC/IFORS/IMACS/IFIP Symposium on Large Scale Systems: Theory and Applications, University of Patras, Greece, July 15 -17, 1998.
10. T. G. Crainic , M. Gendreau, Cooperative Parallel Tabu Search for Capacitated Network Design, Journal of Heuristics, 8, 601-627, 2002.

11. Z. Michalewicz, Genetic Algorithms+Data Structures=Evolution Programs, Springer-Verlag, 1992.

9

Evolvable Fuzzy Hardware for Real-time Embedded Control in Packet Switching

Ju Hui Li, Meng Hiot Lim, and Qi Cao

School of EEE, Block S1, Nanyang Technological University, Singapore 639798, (pg01896341 | emhlim | pg04780942)@ntu.edu.sg

In this chapter, we describe a scheme to realize an *Evolvable Fuzzy Hardware* (EFH) for real-time Packet Switching problem. The common challenges of *Evolvable hardware* (EHW) implementation are issues pertaining to *online adaptation*, *scalability* and *termination of evolution* [1]. The proposed EFH addresses these issues effectively. A very interesting advantage of the proposed EFH is that the system performance can be tuned intuitively through parametric adjustment of the fitness function. This advantage gives the EFH system a very special property that conventional scheduling methods cannot fulfill easily. For the hardware implementation of the EFH, real-time fuzzy inference with high-speed context switching capability is necessary. We address this aspect through implementation based on a context independent *reconfigurable fuzzy inference chip* (RFIC).

9.1 Introduction to EHW and EFH

Evolvable hardware (EHW) is a new type of hardware whose architecture can be evolved to suit the operating environment. In recent years, it has been attracting greater attention from researchers. The idea behind EHW is based on evolutionary algorithm, a methodology to search the solution space to derive the appropriate hardware architecture. EHW can be classified into *extrinsic* and *intrinsic* EHW based on the scheme of evolution used. Extrinsic EHW relies on a simulated evolutionary process independent of the hardware. It may rely on hardware description languages (HDL), C or other programming languages to represent the circuit and then rely on an evolutionary algorithm to evolve the hardware configuration. Only the elite design is downloaded into the reconfigurable device. Intrinsic evolvability means that the evolution and evaluation of solutions are carried out at the hardware level of the EHW system. In principle, intrinsic EHW can modify its own hardware configuration and behavior autonomously. If the environment changes, the behavior or architecture will also change to maintain an acceptable level of system performance. Currently, there has been great progress made for extrinsic type of EHW [2, 3, 4, 5, 6, 7].

There are also research works that focused on intrinsic EHW. In some reported works, the researchers rely on a semi-intrinsic approach. They use software to realize the evolution part and hardware to carry out evaluation of the derived architecture. After the evolution process, the best chromosome is implemented in hardware. This scheme can be called *offline adaptive intrinsic* EHW. Most of the works on intrinsic EHW up to now can be found in [8, 9, 10]. This type of EHW generally has some advantages over extrinsic EHW. Since it carries out the evaluation in hardware, the evaluation process is very fast, and the performance of the elite is not affected by error in the simulation model. Intrinsic EHW is useful for applications that require online and real-time system reconfiguration. However, the implementation of intrinsic EHW still poses significant challenges for such promising areas.

From the perspective of evolution granularity, current EHW can be classified into three types: transistor level, gate level and function level. Among the three, the transistor level represents the lowest level of evolution granularity. This gives the greatest flexibility because transistors are the smallest components of any circuit. Gate level EHW means that logic gates are the smallest configurable components of the EHW [11, 12, 13, 14, 15, 16, 17]. Functional level EHW carries out the evolution of macro units (adder, multiplier, sine, cosine, etc.) implemented on a special type of FPGA [2, 18, 19]. There are many *functional processing units* (FPU) in the FPGA chip. Each FPU can be configured to perform one of the high-level functions such as addition, subtraction, multiplication, division, sine and cosine. The functions and connections of FPUs are configured based on the elite chromosome. Most of the EHW reported can be categorized into one of these three levels. The limitations of these forms of EHW imply that evolutions can only be done extrinsically or in some instances, intrinsically but in an offline adaptive manner.

For the implementation of intrinsic Evolvable and online adaptive EHW, there are three main open issues that need to be addressed [1]. These issues are briefly outlined below.

Online adaptation: This means that the system hardware is required to adapt during the normal operation. Online adaptation is very hard to realize because the system has to reconfigure the hardware for every chromosome in order to carry out the evaluation. Some chromosomes may inevitably result in very poor performance. If these chromosomes are evaluated by reconfiguring the hardware, they may potentially result in some damages or disastrous outcome.

Scalability refers to the extensibility of the scheme to handle more complex architecture or configurations. For a typical EHW, the chromosome length may be hundreds or even thousands of genes for a complicated system. The search space represented by a chromosome may be very big. Hence the search by the *genetic algorithms* (GA) for a good solution in such a big solution space may take a very long time.

Termination of evolution pertains to criteria or conditions for stopping the evolution process. For example, one commonly used criterion is the number of runs. With a GA scheme, there is no guarantee as to the number of runs required before a desirable solution can be found. This can be a significant drawback for real-time operation.

In order to perform online adaptive and intrinsic Evolvable hardware, we propose a new form of EHW that is referred to as *Evolvable Fuzzy Hardware* (EFH). EFH can be viewed as a form of *Evolvable fuzzy system* (EFS) whereby the fuzzy inference system is implemented in hardware to deliver real-time inference throughput. Furthermore, the domain knowledge of the fuzzy system should be able to support online real-time reconfiguration. EFH can overcome the disadvantages of the other three EHWs described earlier and is amenable to intrinsic evolution and online adaptation. Earlier in [20], we proposed EFS for ATM cell scheduling. In that system, the EFS searches for an appropriate fuzzy rule set to carry out the scheduling task on dynamically changing cell flows. The evolutionary search process does not cause any interruption in the system operation. After a good fuzzy rule set is found, the old one is replaced immediately. From simulation results, it was shown that EFS is capable of dynamic real-time adaptation to deliver robust performance. To further support our work, we have also proposed a *reconfigurable fuzzy inference chip* (RFIC) whereby the context can be changed or reconfigured online [21]. By combining the advantages of the EFS and RFIC, we demonstrate in this work how intrinsic Evolvable and online adaptive EFH can be implemented.

In Section 2, we introduce the real-time Packet Switching problem, an application for demonstrating the viability of the EFH. In Section 3, we describe specifically how the implementation challenges of the intrinsic EFH are addressed. In Section 4, we describe the detailed formulation of the fitness function adopted in our EFH. In Section 5, we present the simulation results of applying EFH to solve the real-time problem. Certain desirable properties of the EFH in dealing with the real-time problem are also discussed in this section. In Section 6, we outline details on how the EFH can be implemented from a system's perspective. Finally, we offer some concluding remarks for our work on EFH.

9.2 Packet Switching

Packet Switching is a backbone of modern communication networks. Because of the characteristics of the various services supported by the network, the management of the bandwidth resources is very critical. The multiplexer is an important component used to administer the sharing of bandwidth among different cell flows. It is mainly employed to provide a means of sharing high-speed link for network terminations or network inter-nodes. *Time division scheme* is adopted in the multiplexer. The output link can be divided into different time slots. At anytime, only one input flow is accorded the priority of sending packets through the output channel. The simplified block architecture of the multiplexer is as shown in Fig. 9.1. For illustration, we classify the services into two types, $class_1$ and $class_2$. In the block diagram, BUF_1 and BUF_2 refer to buffers for $class_1$ and $class_2$ respectively. MP represents the time division multiplexing system for transmitting packets through the OUT channel. The *switching control* block is a part of the hardware that handles cell scheduling. When the OUT channel is available, the *switching control* block decides on which cell flow to be sent.

For Packet Switching, $class_1$ can be a form of CBR (*Constant Bit Rate*) traffic, rt-VBR (*real-time Variable Bit Rate*) or both. The $class_2$ traffic type may refer to nrt-VBR (*non-real-time Variable Bit Rate*), UBR (*Unspecified Bit Rate*) or ABR (*Available Bit Rate*) [22]. While $class_1$ type is delay sensitive, $class_2$ is considered to be not sensitive to delay. These two sources of cell flow must be multiplexed on the output channel (OUT) by the MP unit through time division. The capacities of OUT and the input channels are fixed. In this problem, the QoS (*Quality of Service*) of the system can be evaluated by $class_1$ cell delay, $class_2$ cell loss and the balance between $class_1$ cell loss and $class_2$ cell loss. The ideal case is that $class_1$ cell delay and $class_2$ cell loss are very small and there is also a good balance established between $class_1$ cell loss and $class_2$ cell loss.

The application of EHW in ATM cell scheduling has been reported in Liu *et. al* [2, 3]. In their works, the authors presented schemes of functional EHW to solve the problem of cell scheduling. The functional EHW system successfully achieved a circuit that had service performance similar to that of traditional scheduling schemes. However, the scheme has some significant limitations, hence not suitable for practical applications. The main limitation of the system is its inability to evolve intrinsically. Another limitation is that the system had to rely on an external computation platform to carry out evolutionary process due to its large search space. Finally, the system faces the limitation of being trained and tested only on fixed cell flow patterns. In a practical system, the cell flows can change dramatically. There was no effective scheme in this system to adjust the system along with the changing cell flows.

9.3 Solutions for Open Issues

In order to solve the packet scheduling problem, we design the system architecture, incorporating evolutionary mechanisms as in Fig. 9.2. In this system, the training buffers TB_1 and TB_2 are used to store $class_1$ and $class_2$ cells respectively. The size of TB_1 and TB_2 is at least 2 or 3 times that of BUF_1

and BUF_2. When either TB_1 or TB_2 is full, the evolutionary process is triggered. Fitness evaluation is carried out by subjecting each chromosome to the *scheduling model* according to the cell flow stored in TB_1 and TB_2. The purpose of the *scheduling model* is to emulate the function of the multiplexer as in Fig. 9.1. After a specified number of cycles and generations, if a chromosome

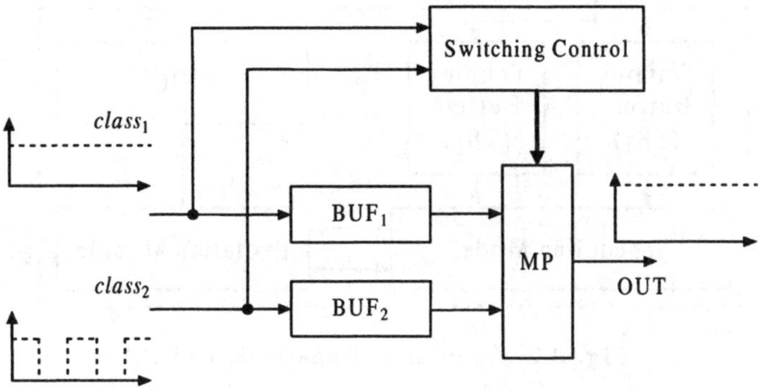

Fig. 9.1. Multiplexer scheme

that corresponds to a system rule set is better than the working chromosome, the working chromosome is replaced immediately. In order to prevent the search procedure from being trapped in a local region, after a pre-specified number of generations, the whole evolutionary process is restarted, from the point where the initial population is generated. This is essentially the start of a new *evolution cycle*. Functionally, the *scheduling model* emulates the packet switching to derive the cell delay and cell loss parameters. This is achieved by a multiplexer model within the *scheduling model* block. The derived parameters enable the fitness value to be calculated using the fitness function. Basically, the *evolution module* evolves the appropriate rule set by interacting with the *scheduling model* to evaluate the fitness of each evolved fuzzy rule set. When evolution is triggered, it works in the background while the MP unit is in operation. With EFH, the fuzzy inference circuit is a very important component and it directly affects the speed of the system's response to the changes in cell flow. Two high-speed fuzzy inference components are required. One is in the *scheduling model* and another is the RFIC block performing cell scheduling control.

During evolution, it is inevitable that poor quality chromosomes i.e., chromosomes that result in poor switching performance, are also evaluated. To avoid the possibility of detrimental effects on the system performance by these chromosomes, the *scheduling model* is incorporated in Fig. 9.2 to emulate the cell scheduling process. This allows for evaluation of the evolved chromosomes in the background. After the evolution process, only the final fuzzy rule set

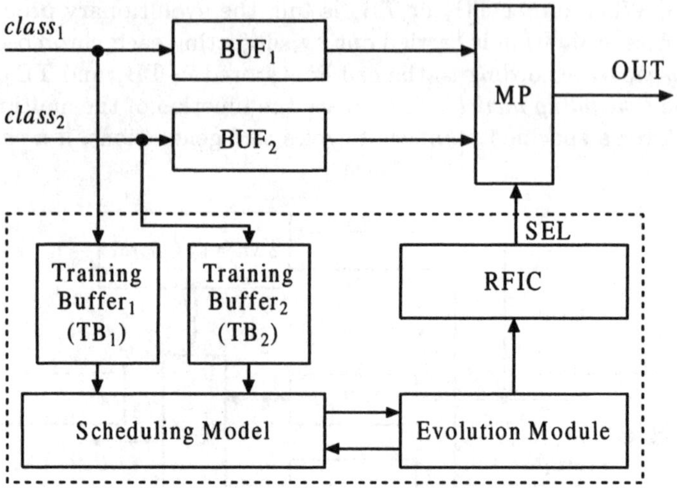

Fig. 9.2. Adaptation framework for EFH

will be configured in the RFIC block. In this way, we address the first major open issue of the intrinsic EHW.

In order to achieve online adaptation and intrinsic evolution for real-time control, another issue that can be regarded as a sub-problem of online adaptation and intrinsic evolution, must also be addressed. During evolution, training data are required. In [2], the EHW system uses the same data for training and testing. This scheme can work well in applications when the real time data do not change dramatically. But if the application scenario is significantly different from the training situation, the system may not perform very well. This indicates that extensive data samples are necessary for such an evolution scheme. If the real-time data change dramatically, it is not practical to incorporate diversely representative real data samples to train the system. For many real-time control areas, we believe that there is no need to do so. In fact, we can apply the principle of "locality" to substantiate this belief. For example, in computer operating system, the design of the cache memory system is based on this principle. Accordingly in computer operating system, if a program is accessing a certain part of the memory, then there is a great likelihood that the program will also access the part of the memory within the same locality in the next time period. In our EFH, we contend that there is a very high probability that the data model within a small time window is the same as the model of data samples in the previous time window. The locality proposition is valid if we assume that the time window is small enough. For the CBR flow, since the cell rate is constant [22, 23], the cell rate at any particular time period is the same as that of the preceding time period. For VBR flow, which can be described by a *two-phase burst/silence* model [2, 24, 25, 26, 27], cells can be sent equidistantly during the burst period and no cells are transmitted during the silence period. The cell rate during the burst period can be approximated based on the principle of "locality". But at the edge of the burst period and the silence period or vice versa, significant error may occur. This kind of prediction error can be tolerated if the time window is sufficiently small. Based on this justification, we can train the system using the previous data flow to approximate the expected data model of the subsequent time period. The smaller the time window, the more flexible the EFH adapts to the cell flow. The best chromosome after an evolution process will be used to do scheduling in the next time period.

To address the scalability issue, we adopt an evolutionary granularity at the fuzzy rule level. In the EFH for Packet Switching, a chromosome can be represented as a string of 25 integers. Each gene of a chromosome represents a fuzzy rule. For this scheme, the search space is not too big compared to the search space in [2, 9], in which each chromosome is represented by a string comprising of hundreds of integers or thousands of bits. The evolution time in the EFH is thus manageable. The third issue to address is the termination of evolution. In many EHW systems, the evolution system may require thousands of generations to get close to an optimal chromosome. The extent of evolution time may limit the applicability of the system for real-time application. In [2], in order to get a good functional EHW to do ATM cell scheduling, the system evolved for 2500 generations with a population size of 400. In [9], in order to derive a circuit with Gaussian output voltage characteristic, the Evolvable hardware system has to evolve 10000 generations. The time scale for evolution in these reported works is not appropriate if used in real-time intrinsic EHW control system. For comparison, in the proposed EFH, a very small population size and small number of the generations are important features of the evolutionary process. In order to prevent the system from adopting a very poor performing fuzzy rule set, we defined a core rule set in the system derived based on the analysis of the problem through human intuition. The core rule set is also used as the startup rule set. If the EFH system is not able to find a chromosome that is better than the core rule set within a fixed number of generations, the core rule set is adopted. The appropriate number of generations for each evolutionary cycle is determined through experimentation. The objective of the evolution is to get a fuzzy rule set better than the working chromosome for the cell flow of the following time period. Even if the derived fuzzy rule set is not optimal, it is deemed to be sufficient. By adopting this idea, the criterion for the termination of evolution can be satisfactorily managed.

9.4 Evolution Scheme

To carry out evolution, GA manipulates a population of chromosomes. These chromosomes are solution representations denoting the application domain fuzzy rule sets,when decoded. In the rest of this section, we will first introduce the fuzzy system and its coding scheme. Then we will describe the inference scheme and the fitness function of this system.

9.4.1 Genetic Coding

A fuzzy system can be formally defined as an application or system, which employs a fuzzy control algorithm. In general, the fuzzy control algorithm refers to a set of *if-then* rules with linguistic values and fuzzy variables. The values are specified as fuzzy concepts defined by membership functions. Fuzzy system implicitly means a set of rules and membership functions.

Suppose a fuzzy system has q input variables x_1, x_2, ..., x_q and single output control variable y, a typical rule for the fuzzy system will be "if $< x_1$ is $A_1 >$ and $< x_2$ is $A_2 >$ ... *and* $< x_q$ is $A_q >$ *then* $< y$ is $D >$". A_1, A_2, ..., A_q and D are fuzzy concepts or linguistic values. Usually, the development of a fuzzy system involves specifying a finite set of labels to represent the linguistic values for describing each of the variables. If the number of labels for the input variables x_1, x_2, ..., x_q are ξ_1, ξ_2, ..., ξ_q respectively, then the number of rules that one can declare will be $\xi_1 \times \xi_2 \times \ldots \times \xi_q$. We refer to this as the maximum or exhaustive rule set. An n-rule fuzzy system would therefore refer to a system with n being less than or equal to $\xi_1 \times \xi_2 \times \ldots \times \xi_q$. This is refered to as an n-rule constrained fuzzy system or simply an n-rule fuzzy system [28, 29, 30].

To begin with, we define two symbols for the inputs, c_1 and c_2. The symbol c_1 refers to the status of $class_1$ cell flow, which is a function of V_1 and V_{max}. V_1 is the current cell rate of $class_1$ cell flow while V_{max} is the line capacity. The symbol c_2 refers to the buffer status of BUF_2. It is a function of L_2 and L_{max}. L_2 is the number of empty units in BUF_2 while L_{max} is the length of BUF_2. For c_1 and c_2, the memberships are characterized by the term set $\{VS, S, M, L, VL\}$ as depicted in Fig. 9.3. These are standard triangular membership functions. The output SEL of the *fuzzy switching control* block (see Fig. 9.2) is characterized by the term set $\{T, F\}$. Both T and F are singletons, or fuzzy sets with impluse membership functions as shown in Fig. 9.4. Functionally, a T or *true* means that the MP unit allocates time slots to cater for the $class_1$ cell flow in BUF_1. An output F or *false* implies that switching is reverted to cells in BUF_2.

Based on the above characterization of the switching network, it is possible to define the n-rule heuristics to control the switching behavior. With the fuzzy memberships defined, one can rely on intuitive logic to define the necessary input-output mappings as shown in Table 9.1. The 25-rule system serves as the default cell scheduling algorithm on system startup. We refer to this rule set as the core rule set.

A fuzzy rule set can be represented as a string of integers. For example, the genetic code for the 25-rule system in Table 9.1 can be described by the string "2221122111221112111111111". The allelic code 1 and 2 correspond to the labels *true* and *false* respectively. The position of the gene in the string

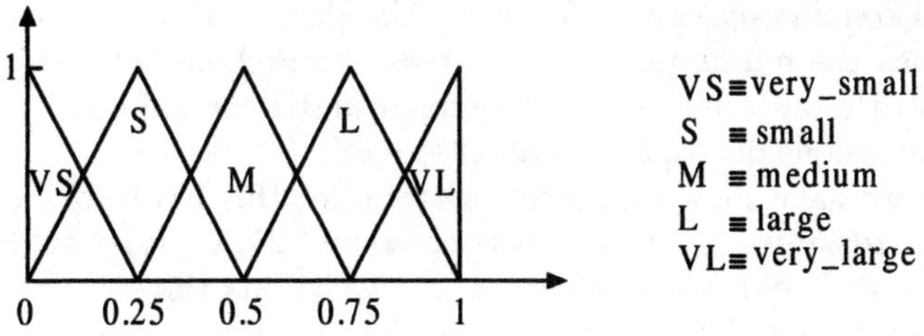

Fig. 9.3. Membership functions for c_1 and c_2

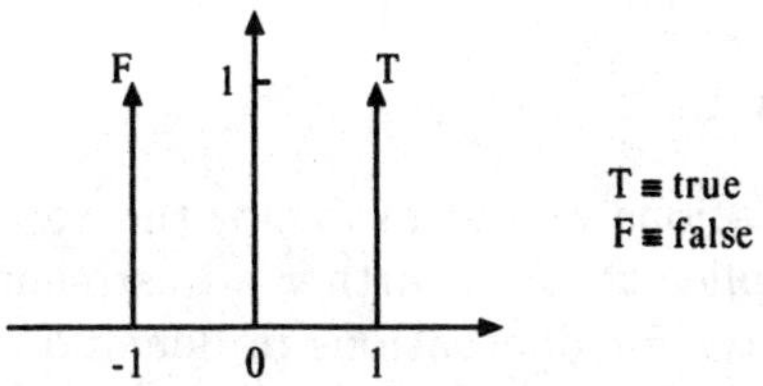

Fig. 9.4. Membership functions for T and F

identifies a specific rule in Table 9.1 when interpreted accordingly in a row wise manner. If the value of a gene is 0, it means that there is no specific fuzzy rule defined for the corresponding input condition. The core rule set not only serves as the startup rule set, but also provides a means to benchmark the performance during the evolution of chromosomes. This scheme guarantees that the performance of the system is better than or at least comparable to that of the core rule set.

Table 9.1. A 25-rule fuzzy system for ATM cell scheduling

Fuzzy Variables		c_1				
		VS	*S*	*M*	*L*	*VL*
c_2	*VS*	*F*	*F*	*F*	*T*	*T*
	S	*F*	*F*	*T*	*T*	*T*
	M	*F*	*F*	*T*	*T*	*T*
	L	*F*	*T*	*T*	*T*	*T*
	VL	*T*	*T*	*T*	*T*	*T*

9.4.2 Inference Scheme

Each entry in Table 9.1 can be interpreted as a statement of the form "if *antecedent*$_1$ and *antecedent*$_2$ then *conclusion*". The *antecedent*$_{\#}$ represents the fuzzy conditions for c_1 or c_2, characterized over the term set $\{VS, S, M,$

L, $VL\}$. The *conclusion* can be T or F. The degree of firing for each fuzzy rule is taken as the minimum of the degrees of matching between the inputs c_1 and c_2 and the antecedents. The aggregation is carried out by averaging the fuzzy conclusions derived from all the rules.

Although we have shown a 25-rule system, for this Evolvable system, the number of the fuzzy rules can vary between 0 and 25. In order to manage the evolution time and reduce the search space, we can fix the size of the rule set to be less than 25 as in [29], so that the evolution time can be managed. This is because the search space for a reduced rule set is more manageable and hence the evolution efficiency can be significantly improved.

9.4.3 Fitness Function

According to the specifications of the problem, the capacity of the output channel is fixed. This implies that no further adjustment on the output capacity can be made to cater for fluctuations in demand. If the bandwidth is not big enough to meet the demand of the two cell flows, servicing $class_1$ cell would mean filling up the $class_2$ buffer and eventually resulting in cell loss for $class_2$. Hence for a specified requirement on the level of cell delay for $class_1$, a certain expected level of $class_2$ cell loss is inevitable. In other words, the $class_2$ cell loss is constrained by the desired level of $class_1$ cell delay that the system is trying to achieve.

There is one main consideration in formulating the fitness function for the EFH. This pertains to the $class_1$ average cell delay. From the above discussion, it is apparent that the level of $class_2$ cell loss is negatively correlated to the average $class_1$ cell delay. Adjusting $class_1$ cell delay will adversely affect the $class_2$ cell loss. Based on these justifications, the fitness function can be described explicitly as in Eq.9.1.

$$F = \kappa - |AveDelay - \lambda \times DelayFactor| \tag{9.1}$$

In Eq.9.1, κ is a very large numerical constant. It is used to adjust the range of fitness values such that F is proportional to the fitness measure of the chromosome. The larger the fitness value, the fitter the chromosome. *AveDelay* is the average delay of $class_1$ cell units after all the cells in TB_1 have been processed. *DelayFactor* is a constant used as a reference for scaling the value of λ based on the desired $class_1$ cell delay. λ is an adjustable coefficient to denote the desired level of average cell delay for $class_1$ cell units stored in TB_1. In general, the system tries to search for a chromosome with minimum $|AveDelay - \lambda \times DelayFactor|$. Both the *AveDelay* and *DelayFactor* in Eq.9.1 can be determined from Eq.9.2 and Eq.9.3 respectively.

$$AveDelay = \frac{1}{\tau} \times \sum_{i=1}^{\tau} m(i) \tag{9.2}$$

$$DelayFactor = \rho \times \upsilon \tag{9.3}$$

In Eq.9.2, $m(\mathrm{i})$ is the waiting time of the ith cell in TB_1 before being sent out. τ is a variable denoting the number of $class_1$ cell units in TB_1 sent during evaluation. $\sum_{i=1}^{\tau} m(i)$ is the sum of the cell delay of the cell units in TB_1. In Eq.9.3, ρ is a constant corresponding to the time required to send a cell through the output channel. The value of ρ depends on the bandwidth capacity of the output channel. The symbol v denotes the size of $\mathrm{TB}_{\#}$. With Eq.9.3, a reference value for the possible delay of $class_1$ cell units can be determined.

9.5 Simulation

In order to demonstrate the viability of the EFH scheme, we carried out simulations of EFH in cell scheduling on two different scenarios. In the simulation, we assume the capacity of the output channel (OUT) and the input channels to be 155.52MHz. The two cell flows are as shown in Fig. 9.5.

For $scenario_1$, $class_1$ is the CBR cell flow with cell bit rate of 155.52MHz. $class_2$ is VBR cell flow, also with a cell bit rate of 155.52MHz. The difference is that the VBR specified has a 2ms ON time period and a 2ms OFF time period. This scenario is a very extreme case used to test the system's controllability. In order to simulate the system performance on a more realistic cell flow, we can adopt $scenario_2$. For $scenario_2$, $class_1$ refers to CBR cell flow with a cell bit rate of 100MHz. $class_2$ is VBR cell flow with unknown random cell bit rate. The minimum cell bit rate for VBR is 55.52MHz while the maximum is 155.52MHz. In these two scenarios, since the sum of the CBR and VBR cell rate is larger than the OUT channel's capacity, cell loss is unavoidable. From

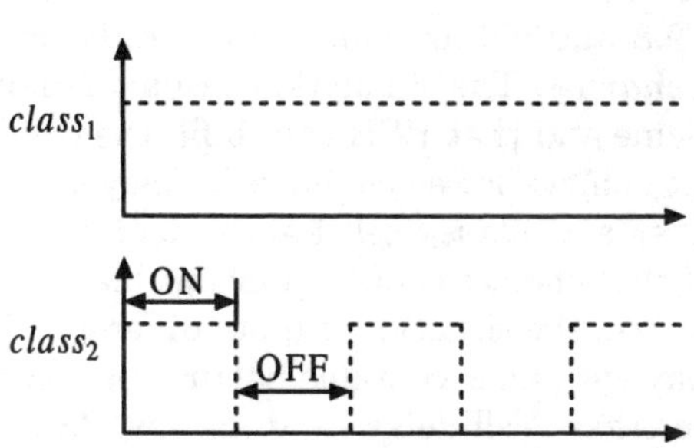

Fig. 9.5. Two classes of cell flows

a practical point of view, the second scenario is more likely compared to the first scenario.

The simulation results are compared with the results of *first-in first-out* (FIFO) and *dynamically weighted priority scheduling* (DWPS) [24]. FIFO is a very traditional scheduling method. It schedules the cell flows based on the arrival time of the packets. FIFO can achieve very good balance between $class_1$ cell loss and $class_2$ loss, but it is very bad in terms of $class_1$ cell delay performance. DWPS is a very good algorithm for cell scheduling. It adjusts the priority according to the cell flow scenarios. But the adaptation scheme of DWPS is not very efficient if the cell flow changes dramatically. DWPS can be described by Eq.9.4. In Eq.9.4, v_i is the fixed priority for different cell flow inputs, a lower value indicates a higher priority. $T_i(t)$ is the waiting time of the oldest packet in the buffer of the ith channel. Q_i is the priority index associated with each cell. The lower the value, the higher the priority. γ is an emphasis parameter and the recommended value is 0.9.

$$Q_i = \frac{v_i}{[T_i(t)]^\gamma} \tag{9.4}$$

9.5.1 Simulation Results

For the simulation, the size of BUF_1 and BUF_2 is 100 cells, and the size of TB_1 and TB_2 is 300 cells. In the fitness function, λ is 0.35. All the simulations are carried out by using a C++ program. The setting for the parameters of the evolutionary algorithm is as follows:

- population size = 10;
- elite pool size = 2;
- crossover probability = 0.6;
- mutation probability = 0.05;
- number of generation = 9;
- number of evolutionary cycle = 2.

We simulated each scheduling scheme for cell flows lasting 2 seconds. Fig. 9.6 and 9.7 are the simulation results of FIFO, DWPS and EFH schemes on $scenario_1$. Fig. 9.8 and 9.9 are simulation results for FIFO, DWPS and EFH schemes on $scenario_2$. The simulation results demonstrate the viability of the evolution scheme and that EFH can fulfill the cell scheduling task. For $scenario_1$, EFH can achieve lower $class_1$ cell delay than FIFO and DWPS. The balance of $class_1$ and $class_2$ cell loss by using these three methods is acceptable. None of the schemes show significant bias towards any of the two cell flows. For $scenario_2$, the situation is quite different. EFH can still achieve lower $class_1$ cell delay with an acceptable balance between $class_1$ cell loss and $class_2$ cell loss. The $class_1$ cell delay by using DWPS is higher than that of EFH and the balance between the $class_1$ cell loss and the $class_2$ cell loss is not good. So according to the quality factors as discussed in Section 9.2, EFH can control the cell scheduling better than FIFO and DWPS when the cell flow changes dramatically.

9.5.2 Tunability of EFH

One advantageous property of EFH is that the system performance can be adjusted very intuitively by decreasing or increasing the value of λ in Eq.9.1. The

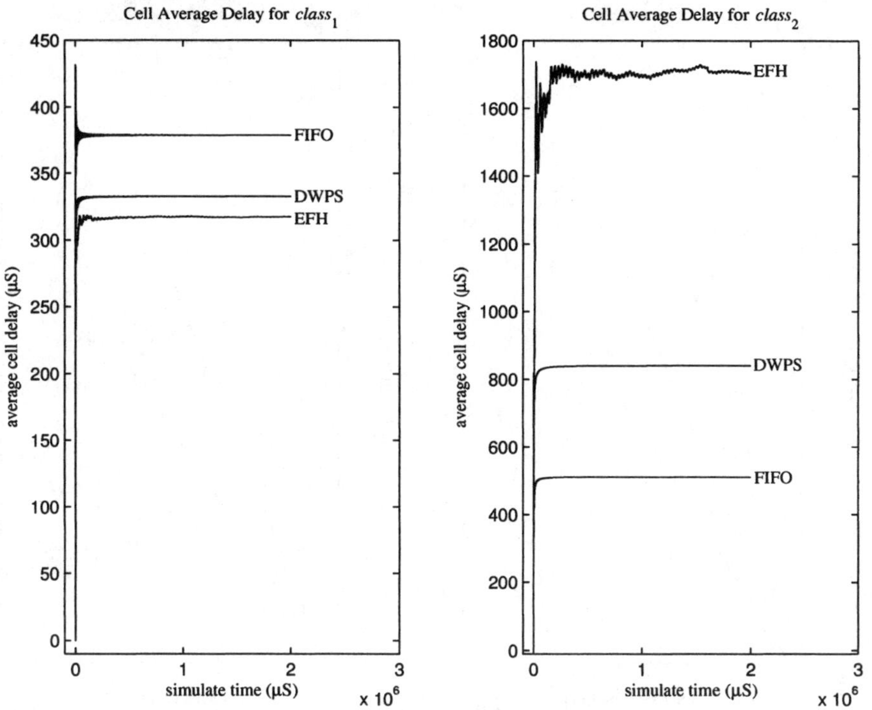

Fig. 9.6. Cell delay for $class_1$ and $class_2$ in $scenario_1$

smaller the value, the smaller the $class_1$ cell delay. This property cannot be achieved conveniently using traditional scheduling methods. As in the above, the tunability of EFH is demonstrated by simulation results on $scenario_1$ and $scenario_2$.

The results of the simulation with different values of λ for $scenario_1$ and $scenario_2$ are as shown in Fig. 9.10, 9.11, 9.12 and 9.13. In Fig. 9.10 and 9.11, when λ is 0.4, the $class_1$ cell delay and $class_1$ cell loss are very small. Accordingly, the $class_2$ cell delay and cell loss are significant. If good balance of $class_1$ cell loss and $class_2$ cell loss is desired, a bigger value can be assigned to λ. In Fig. 9.10 and 9.11, both the $class_1$ cell loss and $class_2$ cell loss are moderate when λ is 0.6. For situations where QoS for $class_2$ needs to be significantly emphasized, the value of λ can be increased. The larger the value for λ, the better the QoS for $class_2$. For example, it is clear from the plots in Fig. 9.10 and 9.11 that λ=0.8 offers good QoS for $class_2$.

For the simulation results in Fig. 9.12 and 9.13 on $scenario_2$, the same conclusion can also be derived. In principle, $class_1$ cell delay can be adjusted in the range from 0 to $\rho \times \upsilon$ if λ is between 0 and 1. This means that $class_1$ cell delay has a very wide range of tunability. It further implies that $class_1$ cell loss and $class_2$ cell loss are also tunable to a wide range. According to the

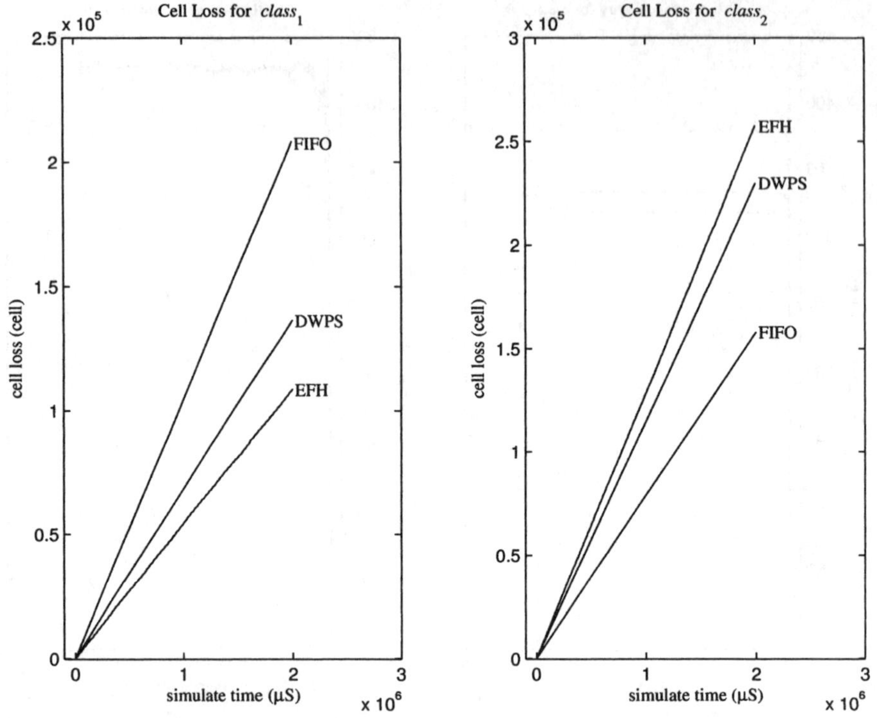

Fig. 9.7. Cell loss for $class_1$ and $class_2$ in $scenario_1$

fitness function, the acceptable level of $class_1$ cell delay can be decided based on the value of λ. On the other hand, if one can decide on the satisfactory $class_1$ cell delay to be achieved, the value of λ can also be approximated.

9.6 Hardware Implementation

According to the evolution scheme described by Fig. 9.2 in Section 9.4, the chromosomes need to be evaluated within a very short time period for each evolution. If the whole evolution process can be completed within the time it takes to send one cell packet through the OUT channel, and a good fuzzy rule set can be found during this time period, the system will enjoy the greatest flexibility in adapting to the changing environment. On the whole, the performance of the system is very much dictated by the quality of the rule set being applied. Each rule set instance is referred to as a context, and is applicable to the current scenario of the operating environment. As context changes, the fuzzy inference circuit is required to accommodate the new context without incurring significant overhead for setup. This implies that a reconfigurable high-speed fuzzy inference circuit is very critical in EFH. In order to achieve

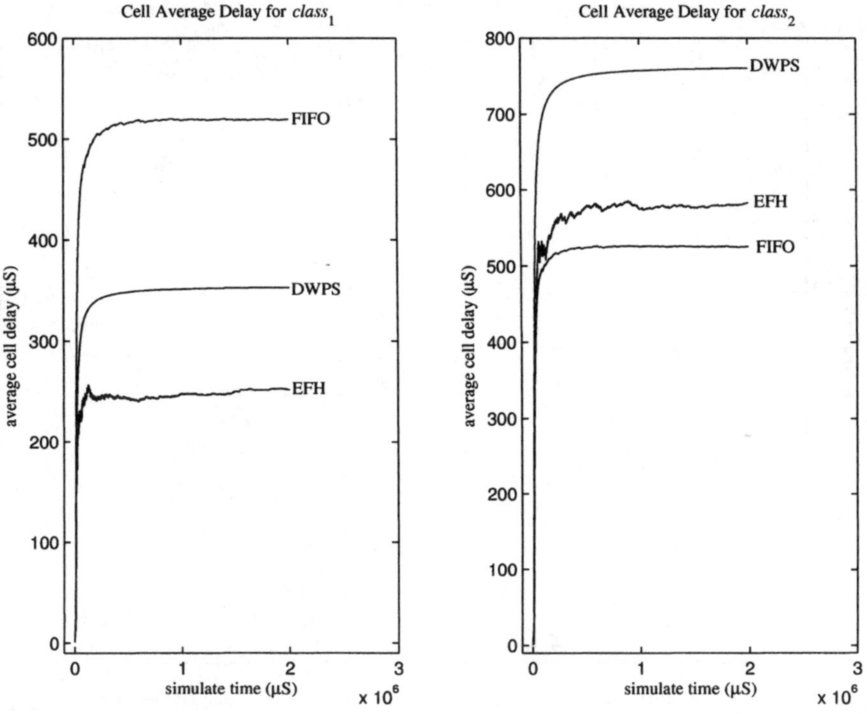

Fig. 9.8. Cell delay for $class_1$ and $class_2$ in $scenario_2$

fast fuzzy inference and at the same time accommodate real-time online context updating, we have proposed a hardware scheme for fuzzy inference called *reconfigurable fuzzy inference chip* (RFIC) [21].

The novelty of the RFIC lies in its ability to accommodate an online context change without interrupting the system operation. The block architecture of RFIC is as shown in Fig. 9.14. The main component is the FIM (*fuzzy inference map*) block. It adopts an implicit inference approach to deliver high inference speed for applications with dynamically changing contexts. The current applicable context is managed by the CMU (*context management unit*). It stores the working fuzzy context and generates control signals such as $Ena_{<x,y>}$ and $Sel_{<x,y>}$ for the FIM. AEM (*address encoding mechanism*) is the module that generates the address to access the FIM partition blocks activated by the $Ena_{<x,y>}$ signals. The OAM (*output aggregation mechanism*) is the dedicated circuit for fuzzy inference aggregation.

The proposed EFH system for cell scheduling is able to accommodate fuzzy rule sets of up to 25 fuzzy rules. Hence, the FIM block incorporates 25 PBs (*partition blocks*); $PB_{<1,1>}$, $PB_{<1,2>}$ ... $PB_{<5,5>}$. Each PB is a mapping that accommodates all the input situations with specific outputs. The mapping for each PB is created based on a software fuzzy inference model. To illustrate

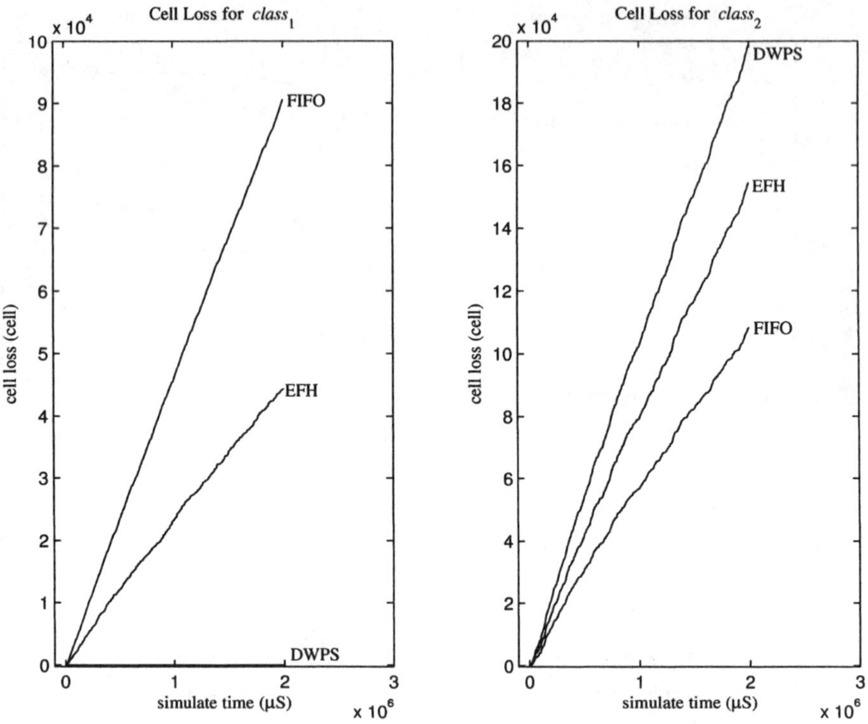

Fig. 9.9. Cell loss for $class_1$ and $class_2$ in $scenario_2$

the basic structure and format of each PB, we can assume that the inputs and the membership functions are digitized to 5 bits. A sample of the mapping data for $PB_{<1,1>}$ is presented in Table 9.2 for illustration. The left column of the table lists the addresses. The whole address string is composed of three parts, i.e., the digitized values of $Input_1$, $Input_2$ and $Sel_{<1,1>}$. The data are made up of two parts. The most significant bit is the fuzzy conclusion bit indicating T or F. The other bits represent the degree of firing for the corresponding fuzzy rule. For example, refering to the first memory unit in Table 9.2, where both $Input_1$ and $Input_2$ equal to "00000", the degree of matching to the membership function *VS* is "11111". So the corresponding datum in the location is "0,11111". Its first bit "0" represents the fuzzy conclusion T and the other bits "11111" is the firing strength.

CMU stores the current application context and generates $Ena_{<x,y>}$ and $Sel_{<x,y>}$ signals. For the application described, the size of the context register required is 50 bits. Each two-bit datum in the register represents a fuzzy rule. The position of each two-bit datum in the 50-bit string identifies the specific rule of the context. A "01" means the fuzzy conclusion is T and "10" indicates the fuzzy conclusion is F. A "00" means that there is no fuzzy rule for the corresponding input situation. Each $Ena_{<x,y>}$ signal can be generated

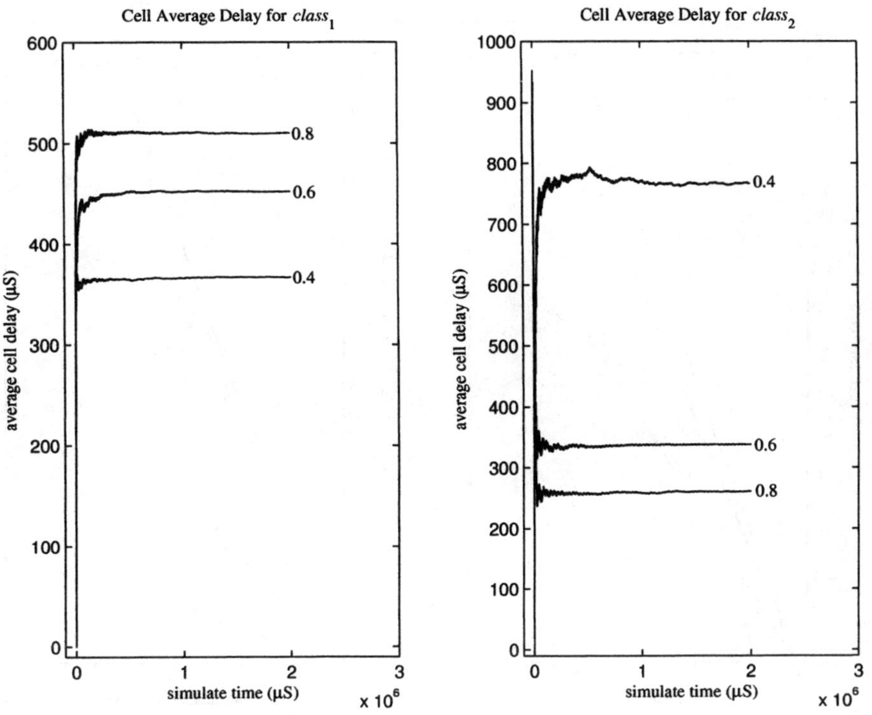

Fig. 9.10. Cell delay for *scenario*$_1$

by applying the logical *OR* operation to the corresponding two bits. A value of "1" for $\mathrm{Ena}_{<x,y>}$ indicates that $\mathrm{PB}_{<x,y>}$ is enabled, which otherwise is disabled. $\mathrm{Sel}_{<x,y>}$ also depends on the specific two bits and is connected to $\mathrm{PB}_{<x,y>}$ separately. A "01" generates a "0" for $\mathrm{Sel}_{<x,y>}$ and "10" produces a "1". The circuit for OAM is as shown in Fig. 9.15. It is made up of Ave_2 blocks and Ave_3 block. In this circuit, the most significant bit of each datum shown in Table 9.2 involve in the aggregration operation is a sign bit. The output has 5 more bits than the input data in order to preserve calculation precision. The control output is derived from the sign bit, i.e, the most significant bit of the OAM output. A positive value indicates that the inference conclusion is T and a negative means the conclusion is F.

9.7 Conclusions

There are several challenges to the application of Evolvable hardware for solving time critical problems. We highlighted three issues, namely *online adaptation*, *scalability* as well as *termination of evolution*. To realize EHW capable of intrinsic online evolution, these issues have to be considered. In this chapter,

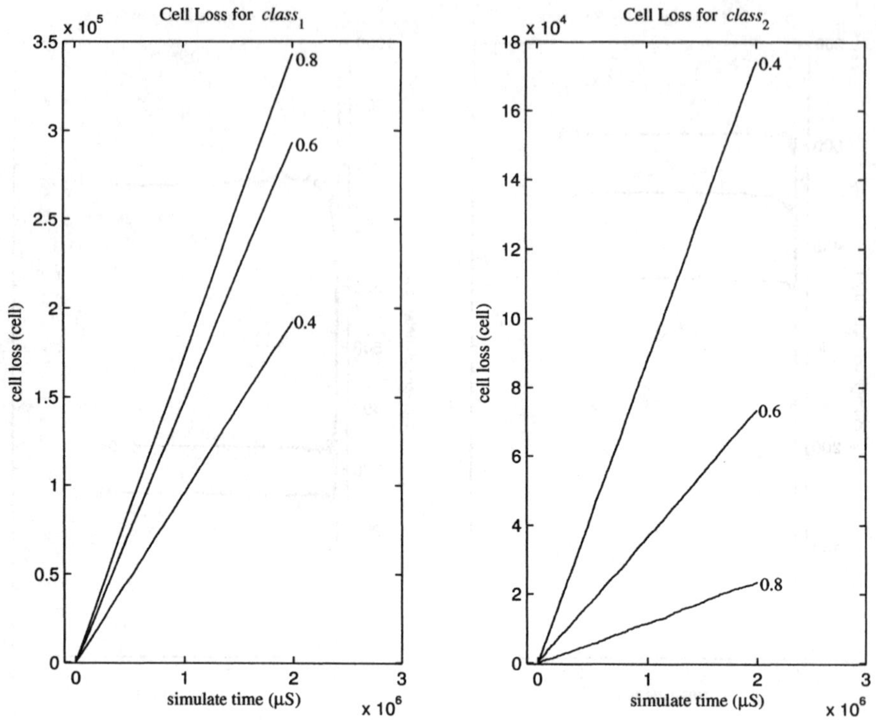

Fig. 9.11. Cell loss for $scenario_1$

Table 9.2. FIM content in $PB_{<1,1>}$

Address	Data
00000,00000,0	0,11111
00000,00000,1	1,11111
00000,00001,0	0,11011
00000,00001,1	1,11011
00000,00010,0	0,10111
00000,00010,1	1,10111
. . .	. . .
00001,00000,0	0,11011
00001,00000,1	1,11011
. . .	. . .
00111,00111,0	0,00111
00111,00111,1	1,00111

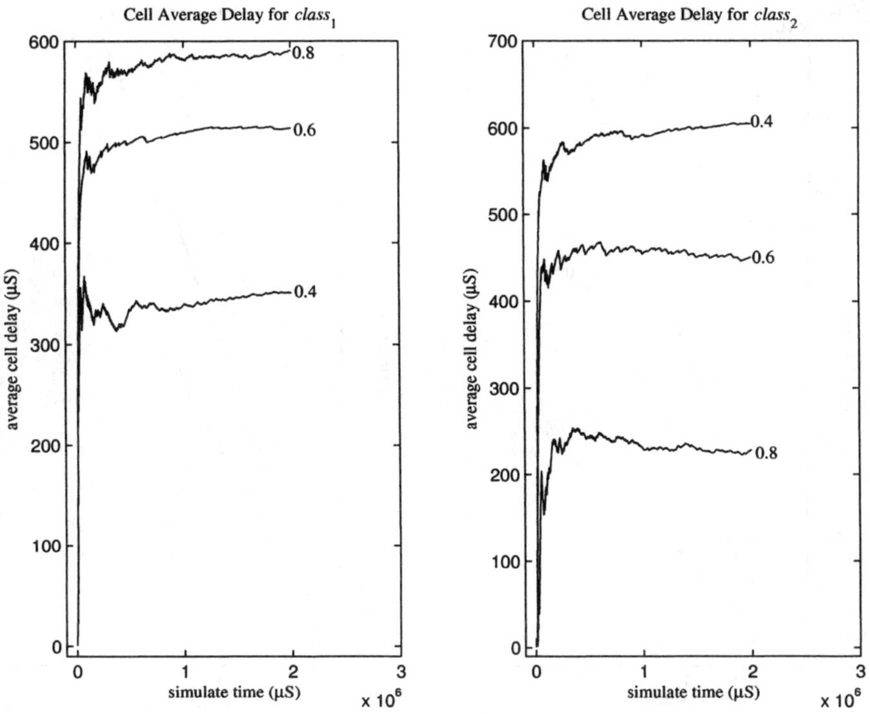

Fig. 9.12. Cell delay for $scenario_2$

we proposed the EFH scheme, a form of EHW whereby the fuzzy inference scheme is carried out in hardware to achieve real-time operation. The scheme allows for updating of online context and domain rules and further incorporating mechanisms to evolve a context appropriate for the application scenario. In order to demonstrate the viability of our proposed EFH, we simulated the control performance of the EFH in cell scheduling and compared the results with some traditional scheduling methods. From the simulation results, it can be seen that the EFH is capable of dealing with changing cell flows much better than the traditional methods. Another significant advantage of the EFH is tunability. This was also analyzed based on the simulation results. Based on analysis of the simulation results, the EFH possesses significant advantages over conventional scheduling methods. To implement the EFH, we described the hardware implementation based on a context switchable RFIC to achieve real-time high-speed fuzzy inferencing and high-speed context updating. By combining this hardware scheme and the evolution scheme, an online adaptive and intrinsic Evolvable EFH can be potentially realized using system-on-chip technology. Although we demonstrated the application of EFH on Packet Switching, the application of EFH is not limited to this. Some real-time control problems such as packet control in parallel computer, token control in

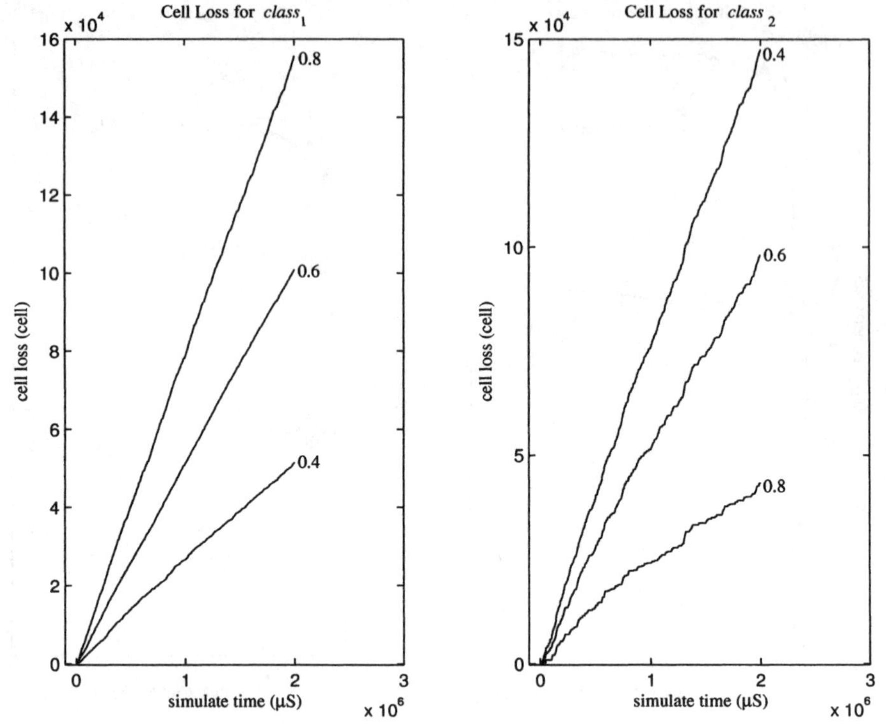

Fig. 9.13. Cell loss for $scenario_2$

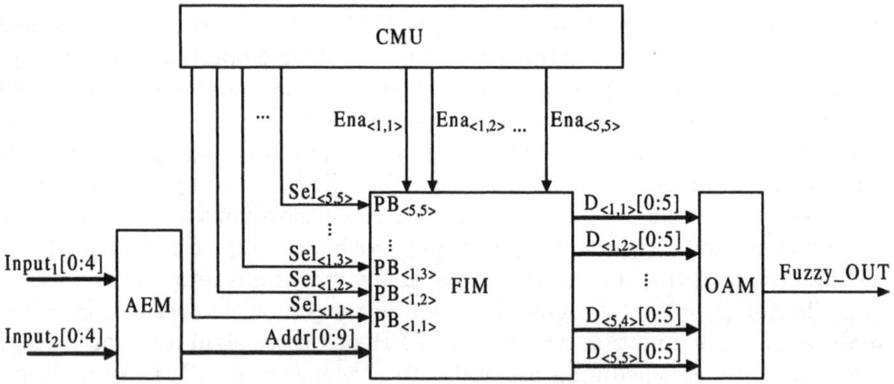

Fig. 9.14. Block architecture of RFIC

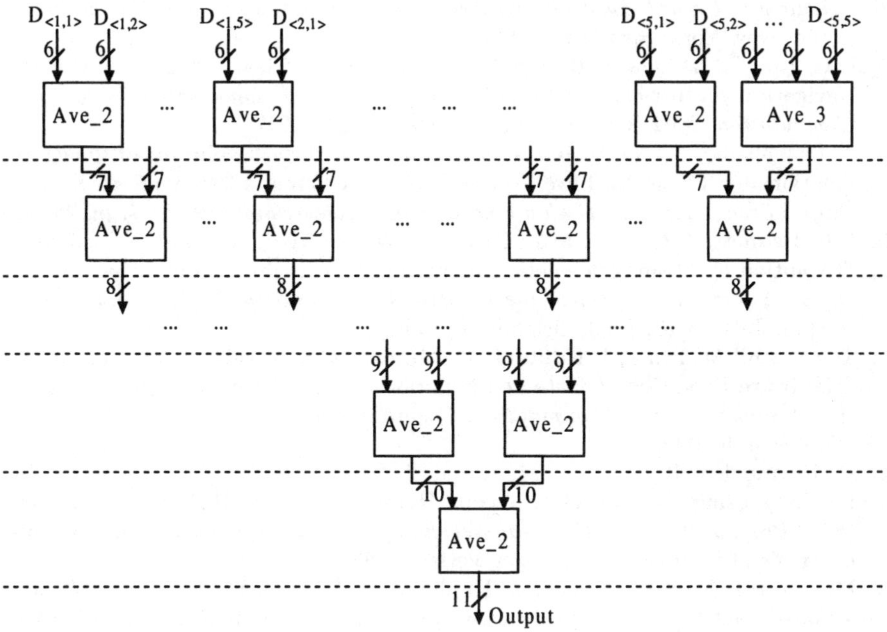

Fig. 9.15. The hardware architecture of OAM

data flow machine, cell flow control in future communication networks are potentially suitable application areas.

References

1. X. Yao and T. Higuchi, "Promises and Challenges of Evolvable Hardware", *IEEE Trans on Syst., Man and Cybern.*, Part C, Applications and Reviews, vol.29, no.1, pp: 87–97, Feb. 1999.
2. W.X. Liu, M. Murakawa and T. Higuchi, "ATM Cell Scheduling by Function Level Evolvable Hardware", LNCS 1259 (ICES1996): pp. 180-192
3. W.X. Liu, M. Murakawa and T. Higuchi, "Evolvable Hardware for On-line Adaptive Traffc Control in ATM Networks", *Genetic Programming 1997, Proc. of the Second Annual Conference*, pp.504–509, Morgan Kaufmann Publishers, 1997.
4. T.G.W. Gordon and P.J. Bentley, "On Evolvable Hardware", In Ovaska, S. and Sztandera, L. (Ed.) *Soft Computing in Industrial Electronics.* Physica-Verlag, Heidelberg, Germany, 2002, pp. 279-323
5. H. D. Garis, "Evolvable Hardware: Principles and Practice", http://www.cs.usu.edu/~degaris/papers/CACM-E-Hard.html
6. M. Iwata, I. Kajitani, H. Yamada, H. Iba and T. Higuchi, "A pattern recognition system using Evolvable hardware", in *Proc. Int. Conf. Parallel Probl. Solving Nature* (PPSN'96).

7. E. Sanchez, *Towards Evolvable hardware: the evolutionary engineering approach,* Berlin; New York: Springer, c1996.
8. K.C. Tan, C.M. Chew, K.K. Tan, L.F Wang and Y.J. Chen, "Autonomous Robot Navigation via Intrinsic Evolution", *Proc. of the 2002 Congress on Evolutionary Computation,* 2002 (CEC'02). vol.2, 2002 pp.1272–1277
9. J. Langeheine, K. Meier and J. Schemmel, "Intrinsic Evolution of Quasi DC Solutions for Transistor Level Analog Electronic Circuits Using a CMOS FPTA Chip". *Proc. NASA/DoD Conference on Evolvable Hardware,* 2002, pp.75–84
10. F.H. Bennett, J.R. Koza, M.A. Keane, J. Yu, W. Mydlowee and O. Stiffelman, "Evolution by Means of Genetic Programming of Analog Circuits that Perform Digital Functions", *Proc. of the Genetic and Evolutionary Computation Conference,* July 13-17, 1999, Orlando, Florida.
11. T. Higuchi, M. Iwata, I. Kajitani, M. Murakawa, S. Yoshizawa and T. Furuya, "Hardware Evolution at Gate and Function Levels," *Proc. Biologically Inspired Autonomous Systems: Computation, Cognition and Action,* Durham, North Carolina, March, 1996.
12. D. Keymeulen, K. Konada, M. Iwata, Y. Kuniyoshi and T. Higuchi, "Robot Learning using Gate-Level Evolvable Hardware", In A. Birk and J. Demiris, (ed.), *Proc. of the Sixth European Workshop on Learning Robots,* Lecture Notes in Artificial Intelligence, Springer-Verlag, 1998.
13. M. Iwata, I. Kajitani, Y. Liu, N. Kajihara and T. Higuchi, "Implementation of a Gate-Level Evolvable Hardware Chip", LNCS 2210 (ICES2001), pp. 38-49, Springer Verlag, 2001.
14. D. Keymeulen, M. Durantez, K. Konaka, Y. Kuniyoshi and T. Higuchi, "An Evolutionary Robot Navigation System using a Gate-Level Evolvable Hardware", LNCS 1259 (ICES1996), pp.195-209, Springer Verlag, 1996.
15. I. Kajitani, T. Hoshino, D. Nishikawa, H. Yokoi, S. Nakaya, T. Yamauchi, T. Inuo, N. Kajihara, M. Iwata, D. Keymeulen and T. Higuchi, "A gate-level EHW chip: Implementing GA operations and reconfigurable hardware on a single LSI", *Evolvable Systems: From Biology to Hardware* (ICES1998), LNCS 1478, pp.1-12, Springer Verlag, 1998.
16. H. Iba, M. Iwata and T. Higuchi, "Gate-Level Evolvable Hardware: Empirical Study and Application", In D. Dasgupta and Z. Michalewicz, editors, *Evolutionary Algorithms in Engineering Applications,* pp.260-275, Springer-Verlag, Berline,1997.
17. H. Iba, M. Iwata and T. Higuchi, "Machine Learning Approach to Gate-Level Evolvable Hardware", *Evolvable Systems: From Biology to Hardware* (ICES1996), LNCS 1259, pp.327-343, Springer-Verlag, 1997.
18. T. Higuchi, M. Murakawa, M. Iwata, I. Kajitani, W. Liu and M. Salami, "Evolvable Hardware at Function Level", *Proc. of 1997 IEEE Int. Conf. on Evolutionary Computation* (ICEC97), pp. 187-192, 1997.
19. M. Murakawa, S. Yoshizawa, I. Kajitani, T. Furuya, M. Iwata and T. Higuchi, "Hardware Evolution at Function Level", *Parallel Problem Solving from Nature–PPSN IV,* LNCS 1141, pp.62-71, Springer-Verlag, 1996.
20. J.H. Li and M.H. Lim, "Evolvable fuzzy system for ATM cell scheduling", *Proc. of 5th Int. Conf. Evolvable Syst.: From Biology to Hardware* (ICES 2003) LNCS 2606, pp. 208-217, Springer-Verlag, 2003.
21. Q. Cao, M.H. Lim and J.H. Li, "A context switchable fuzzy inference chip," submitted to *IEEE Trans. on Fuzzy Syst.*.

22. ATM Forum, "ATM Traffic Management Specification 4.0", April 1996, ftp://ftp.atmforum.com/pub/approved-specs/af-tm-0056.000.pdf
23. R. Jain, "Congestion Control and Traffic Management in ATM Networks: Recent Advances and A Survey," *Computer Networks and ISDN Systems*, vol.28, no.13, October 1996, pp. 1723-1738.
24. T. Lizambri, F. Duran and S. Wakid, "Priority Scheduling and Buffer Management for ATM Traffic Shaping", *Proc. of 7th IEEE Workshop on Future Trends of Distributed Computing Systems, FTDCS'99*, pp.36-43, Dec.20-22, 1999, Cape Town, South Africa.
25. E.P. Rathgeb, "Modeling and Performance Comparison of Policing Mechanisms for ATM Networks", *IEEE J. Select. Areas Commun.*, vol.9, no.3, April 1991.
26. B. Maglaris, D. Anastassiou, P. Sen, G. Karlsson and J.D. Robbins, "Performance models of statistical multiplexing in packet video communications," *IEEE Trans. Commun.*, vol.36, no.7, pp.834-844, July 1988.
27. R. Guerin, H. Ahmadi, M. Naghshineh, "Equivalent Capacity and Its Application to Bandwidth Allocation in High-Speed Networks," *IEEE J. Select. Areas Commun.*, vol.9, no.7, pp968-981, Sept. 1991.
28. M.H. Lim, S. Rahardja and B.H. Gwee, "A GA paradigm for learning fuzzy rules", *Fuzzy Sets and Systems* 82(1996), pp.177-186.
29. M.H. Lim and W.L. Ng, "Iterative Genetic Algorithm for Learning Efficient fuzzy rule Set", to appear in *AIEDAM*, 2004.
30. B. Kosko, *Neural Networks and Fuzzy Systems*, Prentice Hall, Englewood Cliffs, NJ, 1992.

10

Improving Multi Expression Programming: An Ascending Trail from Sea-Level Even-3-Parity Problem to Alpine Even-18-Parity Problem

Mihai Oltean

Department of Computer Science,
Faculty of Mathematics and Computer Science,
Babeş Bolyai University, Kogalniceanu 1, 3400 Cluj-Napoca, Romania,
`moltean@cs.ubbcluj.ro`, `www.cs.ubbcluj.ro/~moltean`

Multi Expression Programming is a Genetic Programming variant that uses a linear representation of individuals. A unique feature of Multi Expression Programming is its ability of storing multiple solutions of a problem in a single chromosome. In this chapter, we propose and use several techniques for improving the search performed by Multi Expression Programming. Some of the most important improvements are Automatically Defined Functions and Sub-Symbolic node representation. Several experiments with Multi Expression Programming are performed in this chapter. Numerical results show that Multi Expression Programming performs very well for the considered test problems.

10.1 Introduction

Multi Expression Programming (MEP)[1] [11, 12, 13] is a new and very efficient technique that may be used for solving difficult real-world problems. A unique feature of MEP is its ability of storing multiple solutions of a problem in a single chromosome. As shown in [11], this feature does not increase the complexity of the decoding process when compared to other Genetic Programming (GP) [7, 8] variants that store a single solution in a chromosome (such as Gene Expression Programming (GEP) [5], Genetic Algorithms for Deriving Software (GADS) [16], Grammatical Evolution (GE) [14], Cartesian Genetic Programming (CGP) [10]).

The MEP technique has been efficiently used for solving symbolic regression problems [11] and even-parity problems [13].

[1] MEP source code is available at www.mep.cs.ubbcluj.ro.

Parity problems arise in many practical applications related to the information technology, especially when data need to be safely transmitted over a network. According to [7] the Boolean even-parity functions are the most difficult Boolean functions to detect via a blind random search. Due to this reason, the ability of the evolutionary algorithms of performing an efficient search in the solutions space can be tested using this problem as a benchmark.

In [13], the MEP has been used for solving even-3 and even-4-parity problems. In this chapter we propose and use several techniques for improving the search performed by Multi Expression Programming. Some of these techniques are:

(i) Automatically Defined Functions (ADFs) [7].
(ii) Sub-Symbolic Node Representation [18].

Numerical experiments performed in this chapter include the use of MEP for solving the even-parity instances from even-3 up to even-18-parity.

MEP without ADFs was able to solve (using a reasonable population and within a reasonable timeframe) up to even-5-parity problem. When Automatically Defined Functions are employed a considerable improvement is obtained, allowing us to evolve a solution for up to even-8-parity problem. More improvements are done when a Sub-Symbolic node representation was employed.

Results of the numerical experiments are compared to those provided by Genetic Programming [7, 8, 18]. It can be easily seen that Multi Expression Programming outperforms Genetic Programming with more than one order of magnitude. Note that a perfect comparison between MEP and GP cannot be made due to the incompatibility of respective representations.

The chapter is organized as follows. In section 10.2 the Even-Parity problem is described. The Multi Expression Programming technique is briefly described in section 10.3. The metrics used to assess the performance of the MEP algorithm are described in section 10.4. Several numerical experiments with MEP for solving the even-3, even-4 and even-5-parity problems are performed in section 10.5. Automatically Defined Functions for MEP are introduced in section 10.6. Several numerical experiments with MEP and ADFs are performed in section 10.7. The sub-symbolic node representation and the smooth operators are introduced in section 10.8. Numerical experiments with MEP and sub-symbolic node representation are performed in section 10.9. Conclusions and the further work directions are suggested in section 10.10.

10.2 Problem Statement

Our aim is to find a Boolean function that satisfies a set of fitness cases. The particular function that we want to find is the Boolean even-parity function. This function has k Boolean arguments and it returns **T** (**True**) if an even number of its arguments are **T**. Otherwise the even-parity function returns **F**

(**False**) [7, 18]. According to [7] the Boolean even-parity functions appear to be the most difficult Boolean functions to detect via a blind random search.

In applying a Genetic Programming technique (particularly Multi Expression Programming) to the even-parity function of k arguments, the terminal set T consists of the k Boolean arguments d_0, d_1, d_2, ... d_{k-1}.

The function set F usually consists of four two-argument primitive Boolean functions (also called gates [9]): AND, OR, NAND, NOR [7, 8]. Using this set we can obtain a solution for small instances of the even-parity problem. Genetic Programming with Automatically Defined Functions has obtained a solution for up to even-11-parity problem using a reasonable population size. If we extend this set by including other Boolean functions (such as EQ and XOR) we can obtain solutions for larger instances. For instance, in [18] Genetic Programming using an extended set of function symbols has been used for solving up to even-22-parity problems. Note that in this case a parallel variant of GP was used on a network of computers structured in a client-server architecture.

The set of fitness cases for this problem consists of the 2^k combinations of the k Boolean arguments. The fitness of an MEP chromosome is the sum, over these 2^k fitness cases, of the Hamming distance (error) between the returned value by the MEP chromosome and the correct value of the Boolean function. Since the standardized fitness ranges between 0 and 2^k, a value closer to zero is better (the fitness is to be minimized).

10.3 Multi Expression Programming

In this section the *Multi Expression Programming* (MEP) [11] paradigm is briefly described.

10.3.1 Individual Representation

MEP genes are represented by substrings of a variable length. The number of genes per chromosome is constant and it defines the length of the chromosome. Each gene encodes a terminal or a function symbol. A gene encoding a function includes references towards the function arguments. Function arguments always have indices of lower values than the position of that function in the chromosome.

This representation is similar to the way in which ***C*** and ***Pascal*** compilers translate mathematical expressions into machine code [1].

MEP representation ensures that no cycle arises while the chromosome is decoded (phenotypically transcripted). According to the representation scheme the first symbol of the chromosome must be a terminal symbol. In this way only syntactically correct programs (MEP individuals) are obtained.

Example. We employ a representation where the numbers on the left positions stand for gene labels (or memory addresses). Labels do not belong to the chromosome, they are provided here only for explanation purposes.

For this example, we use the set of functions $F = \{+, *\}$ and the set of terminals $T = \{a, b, c, d\}$. An example of chromosome using the sets F and T is given below:

1. a
2. b
3. $+$ 1, 2
4. c
5. d
6. $+$ 4, 5
7. $*$ 3, 6

10.3.2 Decoding MEP Chromosome and Fitness Assignment

In this section we described the way in which MEP individuals are translated into computer programs and the way in which the fitness of these programs is computed.

This translation is achieved by reading the chromosome top-down. A terminal symbol specifies a simple expression. A function symbol specifies a complex expression obtained by connecting the operands specified by the argument positions with the current function symbol.

For instance, genes 1, 2, 4 and 5 in the previous example encode simple expressions formed by a single terminal symbol. These expressions are $E_1 = a$, $E_2 = b$, $E_4 = c$ and $E_5 = d$. Gene 3 indicates the operation $+$ on the operands located at positions 1 and 2 of the chromosome. Therefore gene 3 encodes the expression $E_3 = a + b$. Gene 6 indicates the operation $+$ on the operands located at positions 4 and 5. Therefore gene 6 encodes the expression $E_6 = c + d$. Gene 7 indicates the operation $*$ on the operands located at position 3 and 6. Therefore gene 7 encodes the expression $E_7 = (a+b)*(c+d)$, wherein E_7 is the expression encoded by the whole chromosome.

There is neither practical nor theoretical evidence that one of these expressions is better than the others. Moreover Wolpert and McReady [20, 21] proved that we cannot use the search algorithm's behavior so far for a particular test function to predict its future behavior on that function. Thus we cannot choose one of the expressions (let us say expression E_7) to store the output of the chromosome. Even this expression proves to be useful for the first 10 generations we cannot guarantee that it will be the best option for all generations.

This is why each MEP chromosome is allowed to encode a number of expressions equal to the chromosome length. Each of these expressions is considered as being a potential solution of the problem.

The value of these expressions may be computed by reading the chromosome top down. Partial results are computed by Dynamic Programming [2] and are stored in a conventional manner.

As MEP chromosome encodes more than one problem solution, it is interesting to see how the fitness is assigned. Usually the chromosome fitness is defined as the fitness of the best expression encoded by that chromosome. For instance, if we want to solve symbolic regression problems the fitness of each sub-expression E_i may be computed using the formula:

$$f(E_i) = \sum_{k=1}^{n} |o_{k,i} - w_k|,$$

where $o_{k,i}$ is the obtained result by the expression E_i for the fitness case k and w_k is the targeted result for the fitness case k. In this case the fitness needs to be minimized.

The fitness of an individual is set to be equal to the lowest fitness of the expressions encoded in chromosome:

$$f(C) = \min_i f(E_i).$$

When we have to deal with other problems we compute the fitness of each sub-expression encoded in the MEP chromosome and the fitness of the entire individual is given by the fitness of the best expression encoded in that chromosome.

10.3.3 Genetic Operators

Search operators used within MEP algorithm are crossover and mutation. These operators preserve the chromosome structure. All offspring are syntactically correct expressions.

Crossover

By crossover two parents are selected and recombined. For instance, within the uniform recombination the offspring genes are taken randomly from one parent or another.

Example. Let us consider the two parents C_1 and C_2 given in Table 10.1. The two offspring O_1 and O_2 are obtained by uniform recombination as shown in Table 10.1.

Mutation

Each symbol (terminal, function or function pointer) in the chromosome may be the target of mutation operator. By mutation some symbols in the chromosome are changed. To preserve the consistency of the chromosome its first gene must encode a terminal symbol.

Table 10.1. MEP uniform recombination.

Parents		Offspring	
C_1	C_2	O_1	O_2
1: ***b***	1: *a*	1: *a*	1: ***b***
2: *** 1, 1**	2: *b*	2: *** 1, 1**	2: *b*
3: **+ 2, 1**	3: + 1, 2	3: **+ 2, 1**	3: + 1, 2
4: ***a***	4: *c*	4: *c*	4: ***a***
5: *** 3, 2**	5: *d*	5: *** 3, 2**	5: *d*
6: ***a***	6: + 4, 5	6: + 4, 5	6: ***a***
7: **- 1, 4**	7: * 3, 6	7: **- 1, 4**	7: * 3, 6

Example. Consider the chromosome C given in Table 10.2. If the boldfaced symbols are selected for mutation, an offspring O is obtained as given in Table 10.2.

Table 10.2. MEP mutation.

C	O
1: *a*	1: *a*
2: * 1, 1	2: * 1, 1
3: ***b***	3: **+ 1, 2**
4: * 2, 2	4: * 2, 2
5: *b*	5: *b*
6: + **3**, 5	6: + **1**, 5
7: *a*	7: *a*

10.3.4 MEP Algorithm

Standard MEP algorithm uses steady state [19] as its underlying mechanism. MEP algorithm starts by creating a random population of individuals. The following steps are repeated until a given number of generations is reached. Two parents are selected using a selection procedure. The parents are recombined in order to obtain two offspring. The offspring are considered for mutation. The best offspring replaces the worst individual in the current population if the offspring is better than the worst individual. The algorithm returns as its answer the best expression evolved along a fixed number of generations.

10.4 Assessing the Performance of the MEP Algorithm

For assessing the performance of the MEP algorithm three statistics are of high interest:

(i) The relationship between the success rate and the number of genes in a MEP chromosome,
(ii) The relationship between the success rate and the size of the population used by the MEP algorithm,
(iii) The computational effort.

The success rate is computed using the equation (10.1).

$$Success\,rate = \frac{The\,number\,of\,successful\,runs}{The\,total\,number\,of\,runs}. \tag{10.1}$$

Another method used to assess the effectiveness of an algorithm, has been suggested by Koza [7]. The method consists of calculating the number of chromosomes, which would have to be processed to give a certain probability of success. To calculate this figure one must first calculate the cumulative probability of success $P(M, i)$, where M represents the population size, and i the generation number. The value $R(z)$ represents the number of independent runs required for a probability of success (given by z) at generation i. The quantity $I(M, z, i)$ represents the minimum number of chromosomes which must be processed to give a probability of success z, at generation i. The formulae are given by the equations (10.2), (10.3) and (10.4). $Ns(i)$ represents the number of successful runs at generation i, and N_{total}, represents the total number of runs. Note that when $z = 1.0$ the formulae (10.3) and (10.4) are invalid (all runs successful). In the tables and graphs of this chapter z takes the value 0.99.

$$P(M, i) = \frac{Ns(i)}{N_{total}}. \tag{10.2}$$

$$R(z) = ceil\left\{\frac{\log(1-z)}{\log(1-P(M,i))}\right\}. \tag{10.3}$$

$$I(M, i, z) = M \cdot R(z) \cdot i. \tag{10.4}$$

Another important issue is related to the number of function evaluations performed by the considered techniques (MEP and GP in our case). Due to its special Multi-Expression ability MEP performs more function evaluations than GP (considering the same parameters for both algorithms). But, note that 1 function evaluation performed by MEP is not equivalent with 1 function evaluation performed by GP. MEP and GP have the same complexity for the process of decoding the individuals (that is $O(NG)$, where NG is the number of genes). MEP encodes NG solutions in a chromosome whereas GP encodes 1 solution in a chromosome. Thus, the complexity of performing 1 function evaluation is $O(1)$ for MEP and $O(NG)$ for GP. This is why we calculate the computational effort for both MEP and GP using the same formula 10.4 without taking into account the number of genes in a MEP chromosome.

10.5 Numerical Experiments

In this section we perform several experiments with standard MEP for solving several instances of the even-parity problem. General parameter settings for MEP are given in Table 10.3.

Table 10.3. General parameters of the MEP algorithm for solving even-parity problems.

Parameter	Value
Number of generations	51
Mutation probability	0.2
Crossover type	Uniform
Crossover probability	0.9
Selection	q-tournament (q = 10% of the Population size)
Function set	$F = \{$AND, OR, NAND, NOR$\}$

For reducing the chromosome length we keep all the terminals on the first positions of the MEP chromosomes. We also increased the selection pressure by using larger values (usually 10% of the population size) for the tournament sample.

Even-3-parity

The even-3-parity problem has three Boolean inputs and one Boolean output. The number of fitness cases is $2^3 = 8$. The relationship between the success rate and the number of genes in a chromosome and the population size is analyzed for this problem.

A population of 100 individuals has been used when the relationship between the success rate and the chromosome length has been analyzed. Chromosomes of 100 genes have been used for analyzing the relationship between the success rate and the population size. Other parameters of the MEP algorithm are given in Table 10.3. Results are depicted in Fig. 10.1.

Fig. 10.1 shows that MEP is able to solve very well this problem. A population of 240 individuals each having 100 genes (see Fig. 10.1 right side) or a population of 100 individuals with 200 genes (see Fig. 10.1 left side) is sufficient to yield a 100% probability of success GP used [7] a population of 4000 individuals in order to achieve a 100% probability of success for this problem.

The shortest evolved circuit implementing the even-3-parity problem has 6 gates. One of the evolved circuits is depicted in Fig. 10.2.The minimum computational effort required to solve this problem is 6840 and it has been obtained at generation 11 using a population of 40 individuals with 100 genes each.

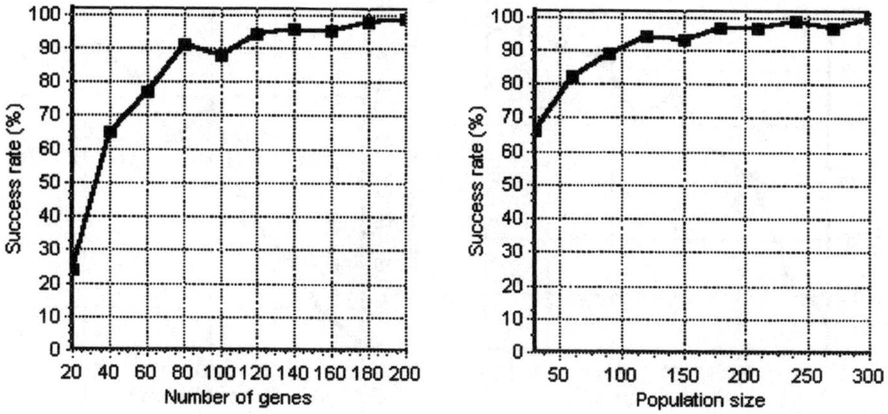

Fig. 10.1. The relationship between the success rate and the chromosome length (left side) and the population size (right side). Results are averaged over 100 runs.

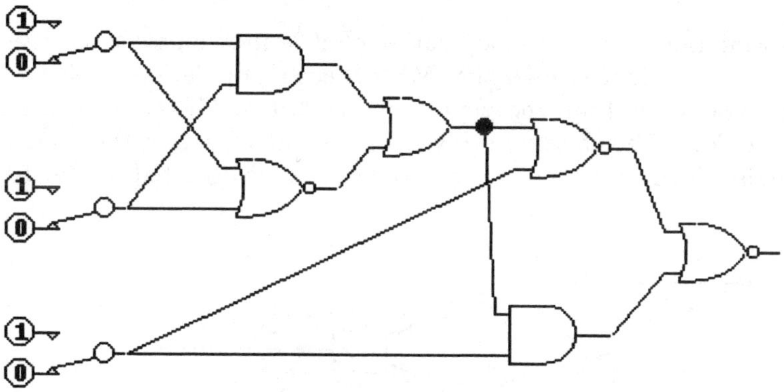

Fig. 10.2. A circuit for the even-3-parity problem.

Even-4-parity

In this experiment, the relationship between the number of genes in a chromosome and the success rate is analyzed for the even-4-parity problem. A population of 400 individuals has been used when the relationship between the success rate and the chromosome length has been analyzed. Chromosomes having 200 genes have been used for analyzing the relationship between the success rate and the population size. Other parameters of the MEP algorithm are given in Table 10.3. Results are depicted in Fig. 10.3.

Fig. 10.3 shows that MEP performs very well on the considered test problem. A population of 200 individuals each having 180 genes is sufficient for yielding a success rate of 42% (see Fig. 10.3 left side).

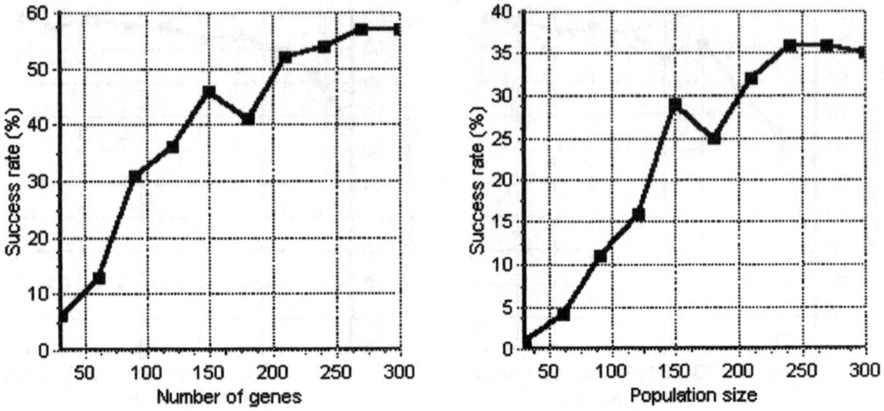

Fig. 10.3. The relationship between the success rate and the chromosome length (left side) and the population size (right side). Results are averaged over 100 runs.

Knowing that GP used a population of 4000 individuals to achieve a success rate of 42% we may infer that MEP needs a population smaller with one order of magnitude than the population needed by GP to solve the even-4-parity problem. The shortest evolved circuit implementing the even-4-parity problem has 9 gates. One of the evolved circuits is depicted in Fig. 10.4.

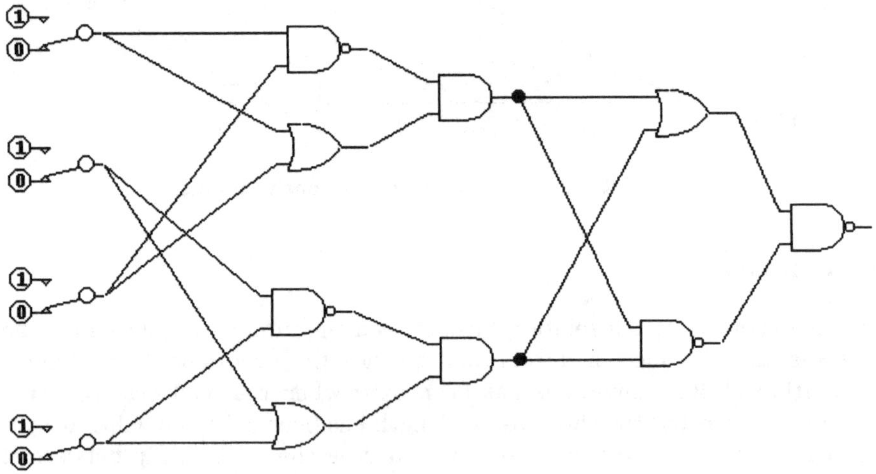

Fig. 10.4. A circuit for the even-4-parity problem.

The minimum computational effort required to solve this problem is 45,900 and it has been obtained at generation 9 using a population of 300 individuals with 200 genes each.

Even-5-parity

In this experiment, the behavior of the MEP algorithm for solving the even-5-parity problem is analyzed. For this problem MEP is run with a population of 4000 individuals having 600 genes each. In 5 runs (out of 30) MEP was able to find a perfect solution for this problem, yielding a success rate of 16.66%.

Note that for this problem GP - without Automatically Defined Functions (ADFs) - was not able to obtain a solution (within 20 runs) with a population of 4000 individuals [7]. When the population size was increased to 8000 individuals a solution was obtained by GP after 8 runs [7].

The curve representing the computational effort needed by MEP to solve the even-5-parity problem is depicted in Fig. 10.5.

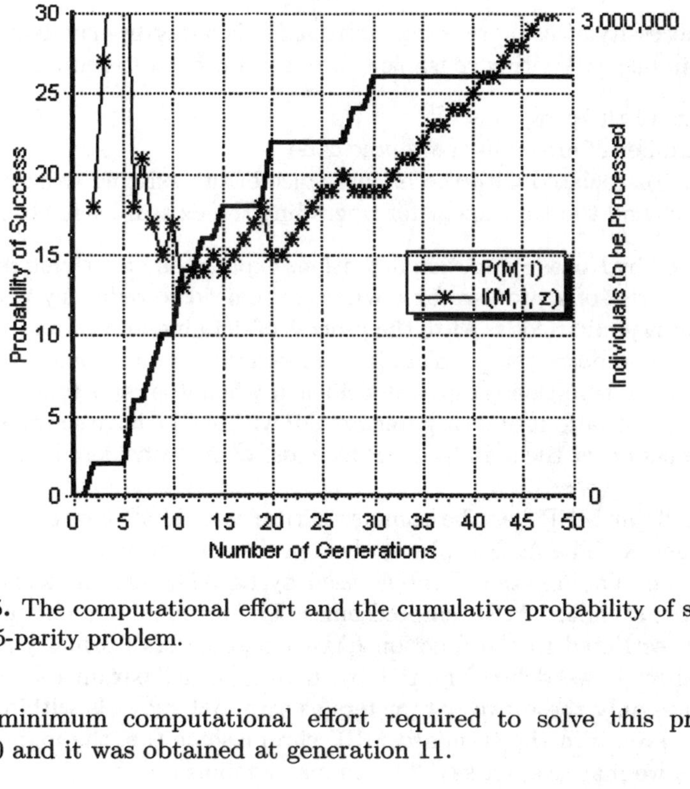

Fig. 10.5. The computational effort and the cumulative probability of success for the even-5-parity problem.

The minimum computational effort required to solve this problem is 1,364,000 and it was obtained at generation 11.

10.5.1 Summarized Results

The results obtained by GP and MEP are summarized in Table 10.4.

Table 10.4 shows that MEP outperforms standard GP with more than one order of magnitude for the even-3 and even-4-parity problems.

We may conclude that MEP significantly outperforms standard GP (without ADFs) for these particular cases of the even-parity problem.

Table 10.4. Computational effort required by GP and MEP for solving several even-parity instances. GP results are taken from [7].

Problem	GP	MEP
even-3-parity	80,000	6,840
even-4-parity	1,276,000	45,900
even-5-parity	6,528,000	1,364,000

10.6 Automatically Defined Functions in MEP

In this section we describe the way in which the Automatically Defined Functions [8] are implemented within the context of Multi Expression Programming.

The necessity of using reusable subroutines is a day-by-day demand of the software industry. Writing reusable subroutines proved to reduce:

(i) the size of the programs.
(ii) the number of errors in the source code.
(iii) the cost associated with the maintenance of the existing software.
(iv) the cost and the time spent for upgrading the existing software.

As noted by Koza [8] function definitions exploit the underlying regularities and symmetries of a problem by obviating the need to tediously rewrite lines of essentially similar code. Also, the process of defining and calling a function, in effect, decomposes the problem into a hierarchy of subproblems.

A function definition is especially efficient when it is repeatedly called with different instantiations of its arguments. GP with ADFs have shown significant improvements over the standard GP for most of the considered test problems [7, 8].

An ADF in MEP has the same structure as a MEP chromosome (i.e. a string of genes). The ADF is also evolved in the same way as a standard MEP chromosome. The function symbols used by an ADF are the same as those used by the standard MEP chromosomes. The terminal symbols used by an ADF are restricted to the function (ADF) parameters (formal parameters). For instance, if we define an ADF with two formal parameters p_0 and p_1 we may use only these two parameters as terminal symbols within the ADF structure, even if in the standard MEP chromosome (i.e. the main evolvable structure) we may use, let say, 20 terminal symbols only.

The set of function symbols of the main MEP structure is enriched with the Automatically Defined Functions considered in the system.

Example. Let us suppose that we want to evolve a problem using 2 ADFs, denoted ADF0 and ADF1 having 2 (p_0 and p_1) respectively 3 (p_0 and p_1 and p_2) arguments. Let us also suppose that the terminal set for the main MEP chromosome is $T = \{a, b\}$ and the function set $F = \{+, -, *, /\}$. The terminal and function symbols that may appear in ADFs and main MEP chromosome are given in Table 10.5.

Table 10.5. Parameters, terminal set and the function set for the ADFs and for the main MEP chromosome.

	Parameters	**Terminal set**	**Function set**
ADF0	p_0, p_1	T={p_0, p_1}	F={+,–,*,/}
ADF1	p_0, p_1, p_2	T={p_0, p_1, p_2}	F={+,–,*,/}
MEP chromosome	–	T={a, b}	F={+,–,*,/, ADF0, ADF1}

The ADF0 (p_0, p_1) could be defined as follows:

1. p_0
2. + 1, 1
3. p_1
4. / 3, 2
5. * 2, 4

The main MEP chromosome could be the following:

1. a
2. b
3. + 1, 2
4. ADF0 3, 1
5. a
6. ADF1 4, 5, 5
7. * 3, 6

The fitness of a MEP chromosome is computed as described in section 10.3.2. The quality of an ADF is computed in a similar manner. The ADF is read once and the partial results are stored in an array (by the means of Dynamic Programming [2]). The best expression encoded in the ADF is chosen to represent the ADF.

The genetic operators (crossover and mutation) used in conjunction with the standard MEP chromosomes may be used for the ADFs too. The probabilities for applying genetic operators are the same for MEP chromosomes and for the Automatically Defined Functions. The crossover operator may be applied only between structures of the same type (that is ADFs having the same parameters or main MEP chromosomes) in order to preserve the chromosome consistency.

10.7 Numerical Experiments with MEP and ADFs

In this section, several numerical experiments with Multi Expression Programming and Automatically Defined Functions are performed. The experiments performed in this section show that the ADF mechanism greatly improves the quality of the search, allowing us to perform a detailed analysis up to the even-8-parity problem. General parameters for Multi Expression Programming are given in Table 10.6.

Table 10.6. The general parameters of MEP with ADFs for solving even-parity problems.

Parameter	Value
Number of generations	51
Mutation probability	0.02
Crossover type	Uniform
Selection	q-tournament (q = 10% of the Population Size)
Function set	F = {AND, OR, NAND, NOR}

All terminals are kept on the first positions of the MEP chromosomes. The tournament size is set to 10% of the population size).

Even-4-parity

In this experiment the relationship between the success rate, the population size and the chromosome length for the even-4-parity problem is analyzed.

A population of 200 individuals is used when the relationship between the success rate and the chromosome length is analyzed. Chromosomes having 200 genes is used for analyzing the relationship between the success rate and the population size. Two Automatically Defined Functions taking two and three arguments are used in conjunction with Multi Expression Programming. The number of genes in ADFs was set to 50. Other parameters are given in Table 10.6. Results are depicted in Fig. 10.6.

The success rate of MEP is 100% when the population size is 200. By contrast, Genetic Programming uses a population of 4000 individuals to obtain the same success rate (100%) [7].

We also computed the effort needed to solve this problem. For this purpose we use a population of 60 MEP individuals having 200 genes each. The number of individuals that needs to be processed in order to obtain a solution with 99% probability is 7,440. This number was obtained at generation 43.

Even-5-parity

For this experiment we use a population with 400 individuals. Each individual has 200 genes. Three Automatically Defined Functions taking two, three and four arguments are used. The number of genes in each ADF is 50. Other MEP parameters are given in Table 10.6.

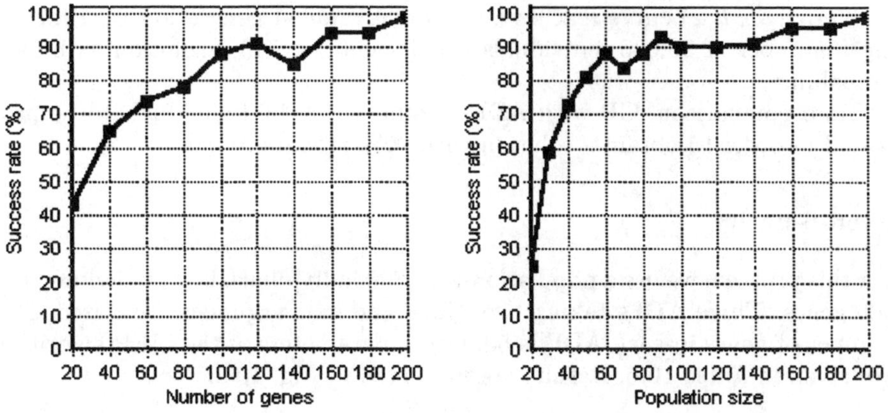

Fig. 10.6. The relationship between the success rate and the chromosome length (left side) and the population size (right side). Results are averaged over 100 runs.

The cumulative probability of success and the computational effort needed for solving this problem are depicted in Fig. 10.7.

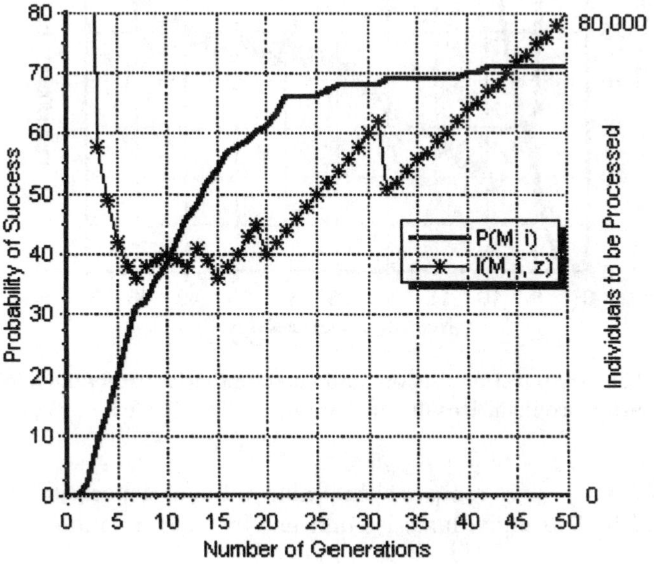

Fig. 10.7. The computational effort and the cumulative probability of success for the even-5-parity problem. Results are averaged over 100 runs.

The $I(M, i, z)$ curve reaches a minimum value at generation 15. Processing a number of 36,000 individuals is sufficient to yield a solution with 99% probability.

As a comparison, GP with ADFs requires 152,000 individuals to be processed in order to obtain a solution with 99% probability [8].

Even-6-parity

For this problem we use a population with 800 individuals. Each individual has 300 genes. Three ADFs taking two, three and four arguments are used. The number of genes in each ADF is 50. Other parameters of the MEP algorithm are given in Table 10.6. Results are presented in Fig. 10.8.

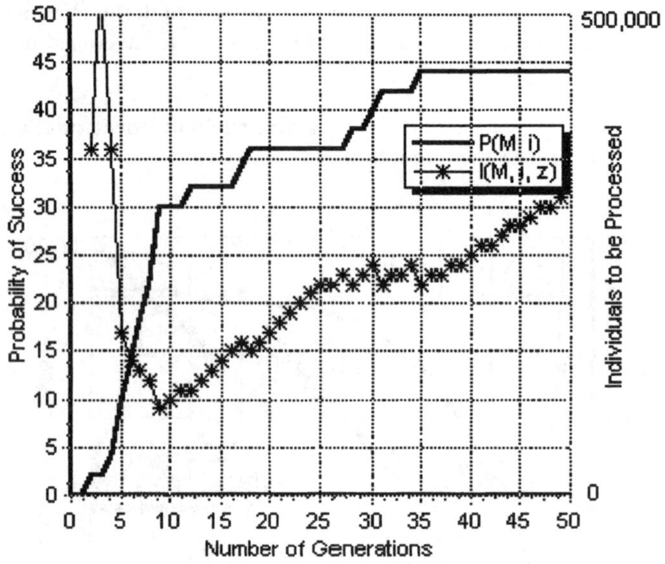

Fig. 10.8. The computational effort and the cumulative probability of success for the even-6-parity problem. Results are averaged over 50 runs.

The $I(M, i, z)$ curve reaches a minimum value at generation 9. Processing a number of 93,600 individuals is sufficient to yield a solution to with 99% probability.

Even-7-parity

For this experiment we use a population with 1000 individuals. Each individual has 400 genes. Three ADFs taking two, three and four arguments are

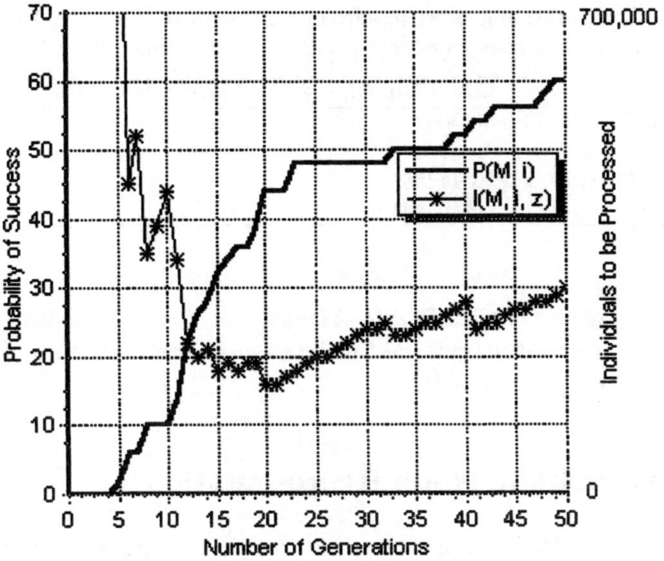

Fig. 10.9. The computational effort and the cumulative probability of success for the even-7-parity problem. Results are averaged over 50 runs.

used. The number of genes in each ADF is 100. Other parameters are given in Table 10.6. Results are given in Fig. 10.9.

Fig. 10.9 shows that the $I(M, i, z)$ curve reaches a minimum value at generation 20. Processing a number of 160,000 individuals is sufficient to yield a solution to with 99% probability. The cumulative probability of success is 60% at generation 50.

Even-8-parity

This case of the even-parity is the most difficult problem analyzed in this section. A population of 1000 individuals is used in this case. Each individual has 400 genes. Three ADFs taking two, three and four arguments are used. The number of genes in each ADF is 100. Other parameters are given in Table 10.6. Due to the increased computational time we performed only five runs which are not sufficient for computing a statistic (i.e. the success rate or the computational effort). A perfect solution (satisfying all fitness cases) was obtained in the fourth run.

10.7.1 Summarized Results

The results obtained by GP and MEP with Automatically Defined Functions are summarized in Table 10.7.

Table 10.7. Computational effort required by GP with ADFs and MEP with ADFs for solving several even-parity instances. GP results are taken from [8].

Problem	GP with ADFs	MEP with ADFs
even-4-parity	80,000	7,440
even-5-parity	464,000	36,000
even-6-parity	1,344,000	93,000
even-7-parity	1,440,000	160,000

Table 10.7 shows that MEP with ADFs outperforms GP with ADFs with more than one order of magnitude for the even-4, even-5, even-6, and even-7-parity problems.

10.8 Sub-Symbolic Node Representation

The Sub-Symbolic Node Representation [15, 18] in order to allow GP to perform small moves in the search space. It is widely known that a single point mutation, that can be applied to a MEP chromosome under the standard representation, may nevertheless result in a significant change in behavior of the MEP program. For instance, consider the gene AND 1 7, where the expressions encoded in positions 1 and 7 are Boolean expressions. If the operator AND is replaced with NAND, the return value of that subtree will be changed for all fitness cases. Instead of such a radical change we want a smoother mechanism that produced a more refined result (that is a mechanism that changes the results produced by only a subset of the training set).

A Boolean function of arity n can be represented as a truth table (bit-string) of length 2^n, specifying its return values on each of the 2^n input combinations. Thus, AND may be represented as 1000, OR as 1110, XOR as 0110. This representation is referred [15, 18] as *sub-symbolic* because function nodes are now seen as collection of entities rather than atomic units.

One feature of the Sub-Symbolic representation of Boolean function nodes is that, in contrast with the reduced function set normally used in Boolean classification tasks, it is unbiased, since it incorporates all 2^n nodes of arity n into its function set. Some of these may be superfluous (e.g. always-ON and always-OFF).

Our principal reason for including all Boolean functions of a given arity in our set is simplicity [18]. IF we want to reduce this set we have to put some constrains in the smooth operators (described in the next section). Note that the EQ and XOR functions are necessarily included in the arity 2 functions sets and that these will probably enhance the performance on the parity problems. On the other hand, the function set is much larger than normal leading to a significantly larger search space.

10.8.1 Smooth MEP Operators

In this section two new MEP operators are proposed. These operators are similar to the standard MEP operators but they can work with the sub-symbolic node representation.

Smooth Uniform Crossover

By crossover two parents are selected and are recombined. For instance, within the uniform recombination the offspring genes are taken randomly from one parent or another. The function parts, which are now binary strings of length 4, are recombined using the uniform crossover from the binary encoding [4].

Example. Let us consider the two parents C_1 and C_2 given in Table 10.8. The two offspring O_1 and O_2 are obtained by uniform recombination as shown in Table 10.8.

Table 10.8. MEP smooth uniform crossover.

Parents		Offspring	
C_1	C_2	O_1	O_2
1: ***b***	1: *a*	1: ***b***	1: *a*
2: **1110 1, 1**	2: *b*	2: *b*	2: **1110 1, 1**
3: **0100 2, 1**	3: 1011 1, 2	3: **0100 2, 1**	3: 1011 1, 2
4: ***a***	4: *c*	4: ***a***	4: *c*
5: **1001 3, 2**	5: *d*	5: *d*	5: **1001 3, 2**
6: ***a***	6: 1111 4, 5	6: 1111 4, 5	6: ***a***
7: **1101 1, 4**	7: 0011 3, 6	7: **1101 1, 4**	7: 0011 3, 6

Smooth Mutation

Each symbol (terminal, function reference and bit encoding the function symbol) in the chromosome may be target of mutation operator. Each binary position encoding the function symbol in a gene is affected by the smooth mutation operator with the same probability as all other symbols in a chromosome. To preserve chromosome consistency its first gene must encode a terminal symbol.

Example. Consider the chromosome C given in Table 10.9. If the boldfaced symbols are selected for mutation an offspring O is obtained as shown in Table 10.9.

Table 10.9. MEP smooth mutation.

C	*O*
1: *a*	1: *a*
2: **1000** 1, 1	2: 1101 1, 1
3: ***b***	3: 1110 2, 1
4: 1101 2, 2	4: 1101 2, 2
5: *b*	5: *b*
6: **1010** **3**, 5	6: 1110 1, 5
7: *a*	7: *a*

10.9 Numerical Experiments with MEP and Sub-Symbolic Representation

The use of Sub-symbolic representation greatly improved the performance of MEP algorithm. Due to this reason we begin our experiments with the even-11-parity problem.

In [18] a parallel version of GP was used to solve the even-parity problem using a sub-symbolic representation. The parallel GP program was run on a client-server architecture with 50 processors. In [18] the authors performed a single run for all instances larger than the even-12-parity problem. More than that, a special technique called sub-machine code GP [17] was used in order to speed-up the GP program. The technique sub-machine code GP make use of processor's ability to perform some operations (such as AND) in parallel for all bits.

Due to the simplicity and efficiency of the MEP algorithm we performed multiple runs (at least 10) for each experiment. This allows us to compute the statistics described in section 10.4. Note that MEP was run on a single processor (at 850 MHz) architecture.

General parameter settings used by MEP in all the experiments performed in this section are given in Table 10.10.

Table 10.10. MEP parameters for solving even-parity problems using a sub-symbolic representation of operators.

Parameter	**Value**
Mutation probability	0.02
Crossover type	Uniform
Crossover probability	0.9
Selection	binary tournament
Function set	16 Boolean functions

Even-11-parity

The even-11-parity problem has 11 Boolean inputs and one Boolean output. The number of fitness cases is $2^{11} = 2048$.

The relationship between the success rate and the number of genes in a chromosome and the population size is analyzed for this problem.

A population of 50 individuals is used when the relationship between the success rate and the chromosome length is analyzed. Chromosomes with 300 genes are used for analyzing the relationship between the success rate and the population size. The number of generations was set to 100. Other parameters of the MEP algorithm are given in Table 10.11. Results are depicted in Fig. 10.10.

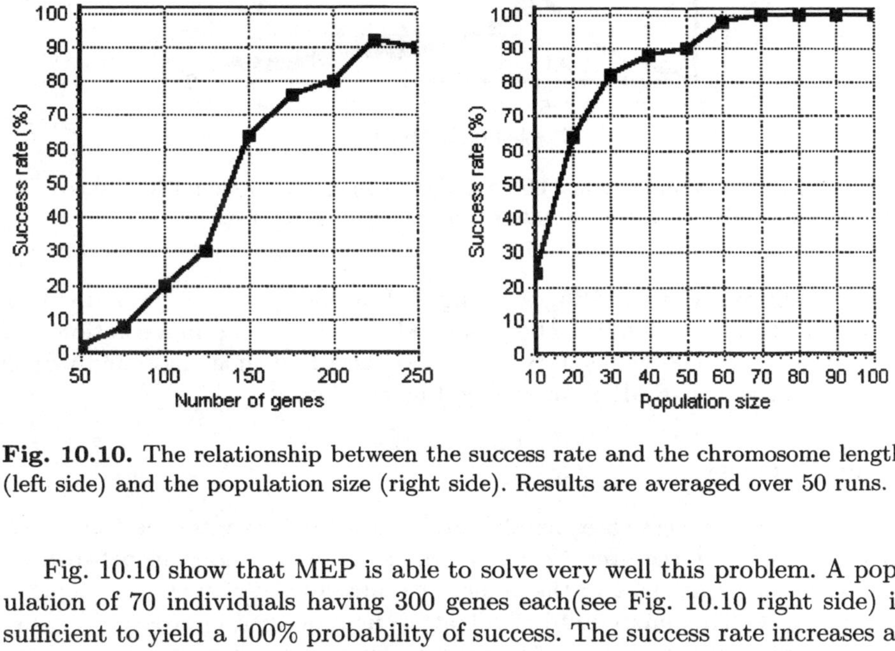

Fig. 10.10. The relationship between the success rate and the chromosome length (left side) and the population size (right side). Results are averaged over 50 runs.

Fig. 10.10 show that MEP is able to solve very well this problem. A population of 70 individuals having 300 genes each(see Fig. 10.10 right side) is sufficient to yield a 100% probability of success. The success rate increases as long as the number of genes in a MEP chromosome increases (see Fig. 10.10).

Even-12-parity

The number of fitness cases for the even-12-parity problem is 4096. For solving this problem with MEP we use a population of 25 individuals having 500 genes each. Other MEP parameters are given in Table 10.10. The program was run for 100 generations. Results over 100 independent runs are presented in Fig. 10.11.

The minimum number of individuals that needs to be processed in order to obtain a solution with a 99% probability of success is 7,420. This number is obtained at generation 99.

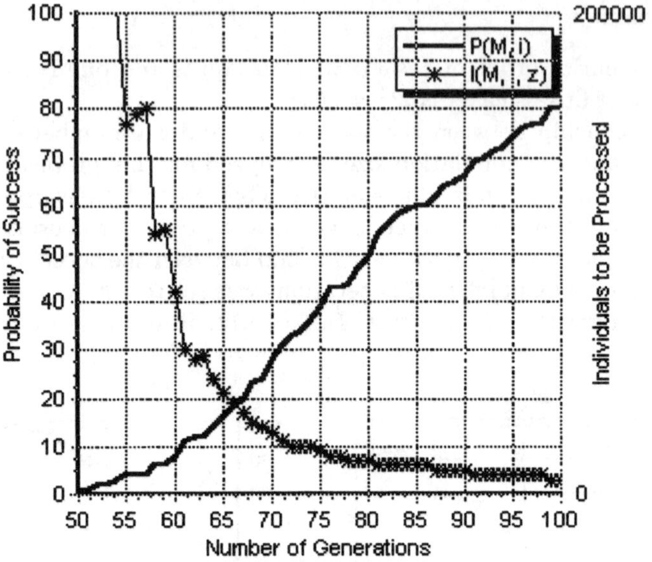

Fig. 10.11. The computational effort and the cumulative probability of success for the even-12-parity problem. Results are averaged over 100 runs.

By contrast, Genetic Programming with a population of 100 individuals requires 98,800 individuals to be processed in order to obtain a solution with 99% probability [18]. Thus, GP requires at least 13.6 times more individuals to be processed than MEP for solving this problem.

Even-13-parity

The number of fitness cases for this problem is 8192. We use the same MEP parameters as for the even-12-parity problem. The relationship between the number of generations and the cumulative probability of success is depicted in Fig. 10.12. The number of individuals to be processed in order to obtain a solution with 99% probability is computed for this problem, too.

The minimum number of individuals that needs to be processed in order to obtain a solution with a 99% probability of success is 2,325. This number is obtained at generation 93.

Even-14-parity

The number of fitness cases for the even-14-parity problem is 16384. For solving this problem with MEP we use a population of 40 individuals having 500 genes each. Other MEP parameters are given in Table 10.10. The program was run for 100 generations. Results over 100 independent runs are presented in Fig. 10.13.

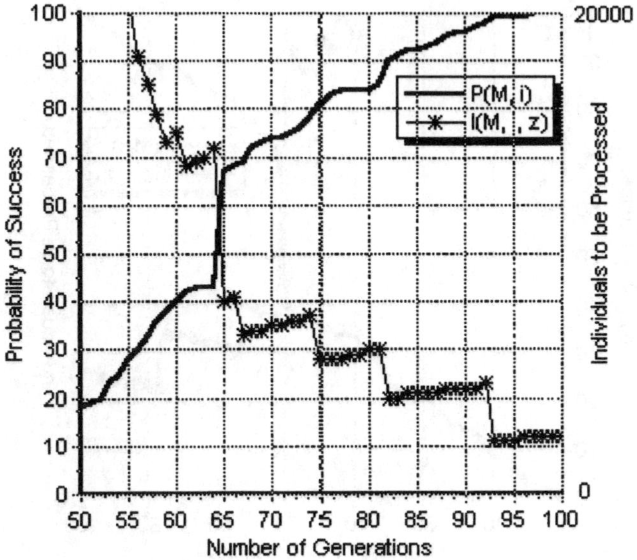

Fig. 10.12. The computational effort and the cumulative probability of success for the even-13-parity problem. Results are averaged over 100 runs.

The minimum number of individuals that needs to be processed in order to obtain a solution with a 99% probability of success is 7,210. This number is obtained at generation 89.

Even-15-parity

The number of fitness cases for the even-15-parity problem is 32768. For solving this problem with MEP we use a population of 100 individuals having 700 genes each. Other MEP parameters are given in Table 10.10. The program was run for 100 generations. Results over 100 independent runs are presented in Fig. 10.14.

The minimum number of individuals that needs to be processed in order to obtain a solution with a 99% probability of success is 29,700. This number is obtained at generation 99.

Even-16-parity

The number of fitness cases for the even-16-parity problem is 65536. For solving this problem with MEP we use a population of 100 individuals having 700 genes each. Other MEP parameters are given in Table 10.10. The program was run for 250 generations. Results over 100 independent runs are presented in Fig. 10.15.

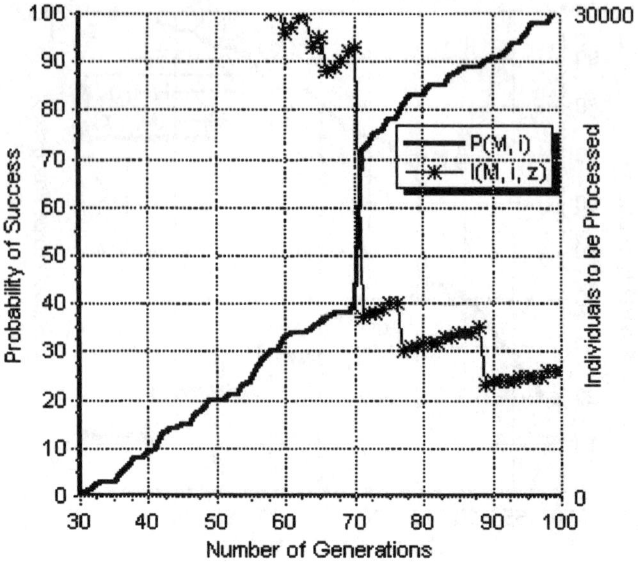

Fig. 10.13. The computational effort and the cumulative probability of success for the even-14-parity problem. Results are averaged over 100 runs.

The minimum number of individuals that needs to be processed in order to obtain a solution with a 99% probability of success is 28,000. This number is obtained at generation 140.

Even-17-parity

For this problem we performed 10 independent runs using the same parameters as those used for the problem even-16-parity. In all runs we obtained a perfect solution. The average number of generations required to obtain a solution is 131.

Even-18-parity

For this problem we performed 6 independent runs using the same parameters as those used for the problem even-16-parity. In 4 runs we obtained a perfect solution. The average number of generations required to obtain a solution is 168.

10.9.1 Summarized Results

The results obtained by MEP with Sub-Symbolic node representation are summarized in Table 10.11.

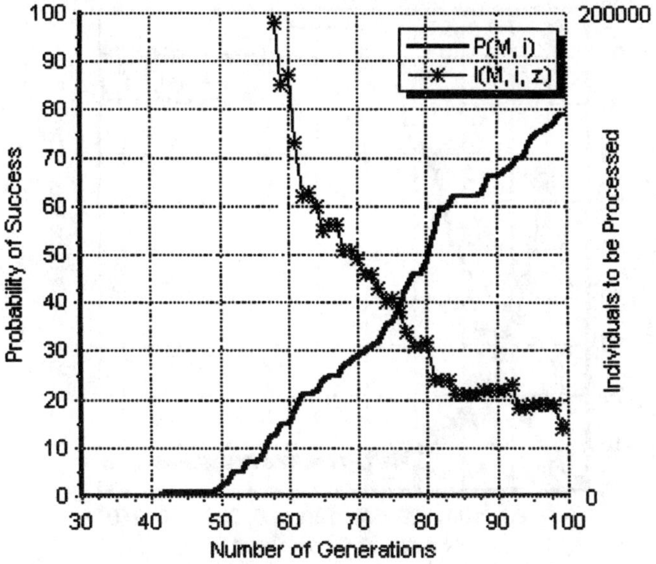

Fig. 10.14. The computational effort and the cumulative probability of success for the even-15-parity problem. Results are averaged over 100 runs.

Table 10.11. Computational effort required by GP and MEP with Sub-symbolic node representation for solving several even-parity instances. GP results are taken from [18].

Problem	**GP with Sub-Symbolic node representation**	**MEP with Sub-Symbolic node representation**
even-12-parity	98,800	7,420
even-13-parity	–	2,325
even-14-parity	–	7,210
even-15-parity	–	29,700
even-16-parity	–	28,000

Table 10.10 shows that MEP is able to solve the considered instances of the parity problem very well. The cells corresponding to GP are empty because GP was run only once for the considered examples.

10.10 Conclusions and Further Work

In this chapter, MEP technique has been used for solving even-parity problems. Two mechanisms for improving the MEP technique have been proposed and tested: Automatically Defined Functions and Sub-symbolic node representation.

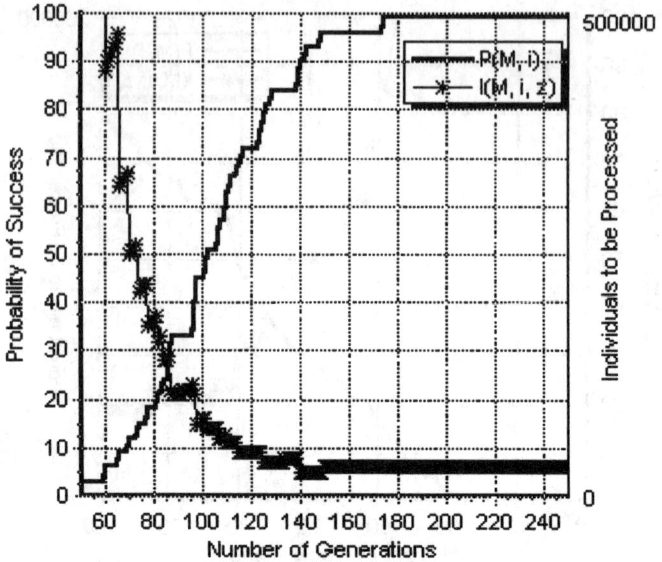

Fig. 10.15. The computational effort and the cumulative probability of success for the even-16-parity problem. Results are averaged over 100 runs.

Tables 10.4, 10.9 and 10.10 show that MEP outperforms GP when the success rate and the number of individuals to be processed is considered. As we said it before this statistics should be interpreted carefully since there are significant differences between GP and MEP representations and a perfect comparison between these two techniques cannot be made.

Further research will be focused on developing a Hierarchically Automatically Defined Functions [8] system within the context of Multi Expression Programming. In this system any function is allowed to call any other function already defined within the system.

Further efforts will be dedicated for implementing a parallel version of MEP (similar to that used in [18] for GP). Using this implementation we will be able to solve other large scale problems including higher versions of the even-parity problem.

Acknowledgments

The author is grateful to anonymous referees for their constructive comments and criticism of earlier versions of this chapter. The title of the chapter is adapted from [6].

References

1. A. Aho, R. Sethi, and J. Ullman, Compilers: Principles, Techniques, and Tools, Addison Wesley, 1986.
2. R. Bellman, Dynamic Programming, Princeton University Press, New Jersey, 1957.
3. M. Brameier, W. Banzhaf, A Comparison of Linear Genetic Programming and Neural Networks in Medical Data Mining, IEEE Transactions on Evolutionary Computation, 5, 17-26, 2001.
4. D. Dumitrescu, B. Lazzerini, L. Jain, A. Dumitrescu, Evolutionary Computation, CRC Press, Boca Raton, FL, 2000.
5. C. Ferreira, Gene Expression Programming: a New Adaptive Algorithm for Solving Problems. Complex Systems, Vol. 13, Nr. 2, pp. 87-129, 2001.
6. A. S. Fraenkel, Scenic trails ascending from sea-level Nim to alpine chess, Games of No Chance, MSRI Workshop on Combinatorial Games, July, 1994, Berkeley, CA, MSRI Publications, R. J. Nowakowski (Editor), Vol. 29, Cambridge University Press, Cambridge, pp. 13-42, 1996.
7. J. R. Koza, Genetic Programming: On the Programming of Computers by Means of Natural Selection, MIT Press, Cambridge, MA, 1992.
8. J. R. Koza, Genetic Programming II: Automatic Discovery of Reusable Programs, MIT Press, Cambridge, MA, 1994.
9. J. Miller, D. Job and V. Vassilev, Principles in the Evolutionary Design of Digital Circuits - Part I, Genetic Programming and Evolvable Machines, Vol. 1, pp. 7 - 35, Kluwer Academic Publishers, 2000.
10. J.F. Miller and P. Thomson, Cartesian Genetic Programming. The 3rd International Conference on Genetic Programming (EuroGP2000), R. Poli, J.F. Miller, W. Banzhaf, W.B. Langdon, J.F. Miller, P. Nordin, T.C. Fogarty (Editors), LNCS 1802, Springer-Verlag, Berlin, pp. 15-17, 2000.
11. M. Oltean and D. Dumitrescu, Multi Expression Programming, technical report, UBB-01-2002, Babes-Bolyai University, Cluj-Napoca, Romania, available at www.mep.cs.ubbcluj.ro, 2002.
12. M. Oltean and C. Groşan, Evolving Evolutionary Algorithms using Multi Expression Programming, The 7th European Conference on Artificial Life, Dortmund, W. Banzhaf (et. al), (Editors), LNCS 2801, pp. 651-658, Springer-Verlag, Berlin, 2003.
13. M. Oltean, Solving Even-parity problems with Multi Expression Programming, The 5th International Workshop on Frontiers in Evolutionary Algorithm, K. Chen (et. al), (Editors) Research Park Triangle, North Carolina, pp. 315-318, 2003.
14. M. O'Neill and C. Ryan, Grammatical Evolution: A Steady State approach, The Second International Workshop on Frontiers in Evolutionary Algorithms, pp. 419-423, 1998.
15. J. Page, R. Poli and W. B. Langdon, Smooth Uniform Crossover with Smooth Point Mutation in Genetic Programming: A Preliminary Study. Genetic Programming, Proceedings of EuroGP'99, R. Poli, P. Nordin, W. B. Langdon and T. C. Fogarty, (Editors), LNCS 1598, pp. 39-49, Springer-Verlag, Berlin, 1999.
16. N.R. Patterson, Genetic Programming with Context-Sensitive Grammars, PhD thesis, University of St. Andrews, Scotland, 2003.

17. R. Poli and W. B. Langdon, Sub-machine Code Genetic Programming, Advances in Genetic Programming 3, L. Spector, W. B. Langdon, U-M O'Reilly and P. Angeline, (Editors), pp. 301-323, MIT Press, Cambridge, MA, 1999.
18. R. Poli and J. Page, Solving High-Order Boolean Parity Problems with Smooth Uniform Crossover, Sub-Machine Code GP and Demes, Journal of Genetic Programming and Evolvable Machines, Kluwer, pp. 1-21, 2000.
19. G. Syswerda, Uniform Crossover in Genetic Algorithms, in Proceedings of the 3rd International Conference on Genetic Algorithms, J.D. Schaffer (Editor), Morgan Kaufmann Publishers, CA, 2-9, 1989.
20. D.H. Wolpert and W.G. McReady, No Free Lunch Theorems for Optimization, IEEE Transaction on Evolutionary Computation, Vol. 1, pp 67-82, 1997.
21. D.H. Wolpert and W.G. McReady, No Free Lunch Theorems for Search, Technical Report, SFI-TR-05-010, Santa Fe Institute, 1995.

Index

Author Index

Reviewer List

Adriane Serapio	John R. Koza
Ajith Abraham	Julian F. Miller
Ali Afzalian	Kalyanmoy Deb
Carlos R. H. Barbosa	Leon Reznik
Carlos A. C. Coello	Luiza M. Mourelle
Cristiana Bentes	Marley M. Vellasco
Dirk Büche	Nadia Nedjah
El-Ghazali Talbi	Orlando Bernardo Filho
Evaristo C. Biscaia Jr.	Peter Dittrich
Felipe G. M. França	Phillip A. Laplante
Flávio J. Souza	Radu-Emil Precup
Gregory Hornby	Ricardo S. Zebulum
Hitoshi Iba	Ricardo Tansheit
Ismat Beg	Saeid Abbasbandy
Janusz Kacprzyk	Tapabrata Ray
Joo A. Vasconcelos	Tim Hendtlass
Johan Andersson	Wolfgang Banzhaf

Printing: Strauss GmbH, Mörlenbach
Binding: Schäffer, Grünstadt